BIOMEDICAL
DIGITAL
SIGNAL
PROCESSING

BIOMEDICAL DIGITAL SIGNAL PROCESSING

**C-Language Examples
and Laboratory Experiments
for the IBM® PC**

WILLIS J. TOMPKINS
Editor

University of Wisconsin-Madison

PRENTICE HALL, *Englewood Cliffs, New Jersey 07632*

Library of Congress Cataloging-in-Publication Data

Biomedical digital signal processing : C-language examples and
    laboratory experiments for the IBM PC / Willis J. Tompkins, editor.
        p.   cm.
    Includes bibliographical references and index.
    ISBN 0-13-067216-5
    1. Signal processing--Digital Techniques. 2. Computers in
medicine. 3. Digital filters (Mathematics) 4. IBM Personal
Computer. 5. C (Computer programming language) I. Tompkins.
Willis J.
R857.D47B56   1993
610'.285--dc20   92-40227
                                                            CIP

Editorial production: *bookworks*   **Editor-in-Chief:** *Bernard Goodwin*
Acquisition editor: *Karen Gettman*   **Prepress buyer:** *Mary McCartney*
Cover designer: *Jeannette Jacobs*   **Manufacturing buyer:** *Mary McCartney*

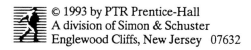

© 1993 by PTR Prentice-Hall
A division of Simon & Schuster
Englewood Cliffs, New Jersey   07632

The publisher offers discounts on this book when ordered in bulk quantities.
For more information write:
    Special Sales/College Marketing
    Prentice Hall
    College Technical and Reference Division
    Englewood Cliffs, New Jersey 07632

Printed in the United States of America

10 9 8 7 6 5 4 3 2

ISBN   0-13-067216-5

Prentice-Hall International (UK) Limited, *London*
Prentice-Hall of Australia Pty. Limited, *Sydney*
Prentice-Hall Canada Inc., *Toronto*
Prentice-Hall Hispanoamericana, S.A., *Mexico*
Prentice-Hall of India Private Limited, *New Delhi*
Prentice-Hall of Japan, Inc., *Tokyo*
Simon & Schuster of Southeast Asia Pte. Ltd., *Singapore*
Editora Prentice-Hall do Brasil, Ltda., *Rio de Janeiro*

Contents

List of Contributors

Valtino X. Afonso
David J. Beebe
Annie P. Foong
Kok Fung Lai
Danial J. Neebel
Jesse D. Olson
Dorin Panescu
Jon D. Pfeffer
Pradeep M. Tagare
Steven J. Tang
Thomas Y. Yen
Ren Zhou

Preface

There are many digital filtering and pattern recognition algorithms used in processing biomedical signals. In a medical instrument, a set of these algorithms typically must operate in real time. For example, an intensive care unit computer must acquire the electrocardiogram of a patient, recognize key features of the signal, determine if the signal is normal or abnormal, and report this information in a timely fashion.

In a typical biomedical engineering course, students have limited real-world design opportunity. Here at the University of Wisconsin-Madison, for example, the laboratory setups were designed a number of years ago around single board computers. These setups run real-time digital signal processing algorithms, and the students can analyze the operation of these algorithms. However, the hardware can be programmed only in the assembly language of the embedded microprocessor and is not easily reprogrammed to implement new algorithms. Thus, a student has limited opportunity to design and program a processing algorithm.

In general, students in electrical engineering have very limited opportunity to have hands-on access to the operation of digital filters. In our current digital filtering courses, the filters are designed noninteractively. Students typically do not have the opportunity to implement the filters in hardware and observe real-time performance. This approach certainly does not provide the student with significant understanding of the design constraints of such filters nor their actual performance characteristics. Thus, the concepts developed here are adaptable to other areas of electrical engineering in addition to the biomedical area, such as in signal processing courses.

We developed this book with its set of laboratories and special software to provide a mechanism for anyone interested in biomedical signal processing to study the field without requiring any other instrument except an IBM PC or compatible. For those who have signal conversion hardware, we include procedures that will provide true hands-on laboratory experiences.

We include in this book the basics of digital signal processing for biomedical applications and also C-language programs for designing and implementing simple digital filters. All examples are written in the Turbo C (Borland) programming language. We chose the C language because our previous approaches have had limited flexibility due to the required knowledge of assembly language programming. The relationship between a signal processing algorithm and its assembly language implementation is conceptually difficult. Use of the high-level C language permits students to better understand the relationship between the filter and the program that implements it.

In this book, we provide a set of laboratory experiments that can be completed using either an actual analog-to-digital converter or a virtual (i.e., simulated with software) converter. In this way, the experiments may be done in either a fully instrumented laboratory or on almost any IBM PC or compatible. The only restrictions on the PC are that it must have at least 640 kbytes of RAM and VGA (or EGA or monochrome) graphics. This graphics requirement is to provide for a high resolution environment for visualizing the results of signal processing. For some applications, particularly frequency domain processing, a math coprocessor is useful to speed up the computations, but it is optional.

The floppy disk provided with the book includes the special program called UW DigiScope which provides an environment in which the student can do the lab experiments, design digital filters of different types, and visualize the results of processing on the display. This program supports the virtual signal conversion device as well as two different physical signal conversion devices. The physical devices are the Real Time Devices ADA2100 signal conversion board which plugs into the internal PC bus and the Motorola 68HC11EVB board (chosen because students can obtain their own device for less than $100). This board sits outside the PC and connects through a serial port. The virtual device simulates signal conversion with software and reads data files for its input waveforms. Also on the floppy disk are some examples of sampled signals and a program that permits users to put their own data in a format that is readable by UW DigiScope. We hope that this program and the standard file format will stimulate the sharing of biomedical signal files by our readers.

The book begins in Chapter 1 with an overview of the field of computers in medicine, including a historical review of the evolution of the technologies important to this field. Chapter 2 reviews the field of electrocardiography since the electrocardiogram (ECG) is the model biomedical signal used for demonstrating the digital signal processing techniques throughout the book. The laboratory in this chapter teaches the student about analog signal acquisition and preprocessing by building circuitry for amplifying his/her own ECG. Chapter 3 provides a traditional review of signal conversion techniques and provides a lab that gives insight into the techniques and potential pitfalls of digital signal acquisition.

Chapters 4, 5, and 6 cover the principles of digital signal processing found in most texts on this subject but use a conceptual approach to the subject as opposed to the traditional equation-laden theoretical approach that is difficult for many students. The intent is to get the students involved in the design process as quickly as possible with minimal reliance on proving the theorems that form the foundation of the techniques. Two labs in these chapters give the students hands-on experience in designing and running digital filters using the UW DigiScope software platform.

Chapter 7 covers a special class of integer coefficient filters that are particularly useful for real-time signal processing because these filters do not require floating-point operations. This topic is not included in most digital signal processing books. A lab helps to develop the student's understanding of these filters through a design exercise.

Chapter 8 introduces adaptive filters that continuously *learn* the characteristics of their processing environment and change their filtering characteristics to optimally perform their intended functions. Chapter 9 reviews the technique and application of signal averaging.

Chapter 10 covers data reduction techniques, which are important for reducing the amount of data that must be stored, processed, or transmitted. The ability to compress signals into smaller file sizes is becoming more important as more signals are being archived and handled digitally. A lab provides an experience in data reduction and reconstruction of signals using techniques provided in the book.

Chapter 11 summarizes additional important techniques for signal processing with emphasis on frequency domain techniques and illustrates frequency analysis of the ECG with a special lab.

Chapter 12 presents a diversity of techniques for detecting the principal features of the ECG with emphasis on real-time algorithms that are demonstrated with a lab. Then Chapter 13 shows how many of these techniques are used in actual medical monitoring systems.

Chapter 14 concludes with a summary of the emerging integrated circuit technologies for digital signal processing with a look to the trends in this field for the future.

Appendices A, B, and C provide details of interfacing and use of the two physical signal conversion devices supported as well as the virtual signal conversion device that can be used on any PC.

Appendix D is the user's manual for the special UW DigiScope program that is used for the lab experiments. Appendix E describes the signal generator function in UW DigiScope that lets the student create simulated signals with controlled noise and sampling rates and store them in disk files for processing by the virtual signal conversion device.

Appendix F covers special problems that can occur due to the finite length of a computer's internal registers when implementing digital signal processing algorithms.

Appendix G reviews some of the commercial software in the market that facilitates digital signal acquisition and processing.

I would especially like to thank the students who attended my class, Computers in Medicine (ECE 463), in Fall 1991 for helping to find many small (and some large problems) in the text material and in the software. I would also like to particularly thank Jesse Olson, author of Chapter 5, for going the extra mile to ensure that the UW DigiScope program is a reliable teaching tool.

The interfaces and algorithms in the labs emphasize real-time digital signal processing techniques that are different from the off-line approaches to digital signal processing taught in most digital signal processing courses. We hope that the use of

PC-based real-time signal processing workstations will greatly enhance the student's hands-on design experience.

Willis J. Tompkins
Department of Electrical
    and Computer Engineering
University of Wisconsin-Madison

BIOMEDICAL DIGITAL SIGNAL PROCESSING

1

Introduction to Computers in Medicine

Willis J. Tompkins

The field of computers in medicine is quite broad. We can only cover a small part of it in this book. We choose to emphasize the importance of real-time signal processing in medical instrumentation. This chapter discusses the nature of medical data, the general characteristics of a medical instrument, and the field of medicine itself. We then go on to review the history of the microprocessor-based system because of the importance of the microprocessor in the design of modern medical instruments. We then give some examples of medical instruments in which the microprocessor has played a key role and in some cases has even empowered us to develop new instruments that were not possible before. The chapter ends with a discussion of software design and the role of the personal computer in development of medical instruments.

1.1 CHARACTERISTICS OF MEDICAL DATA

Figure 1.1 shows the three basic types of data that must be acquired, manipulated, and archived in the hospital. Alphanumeric data include the patient's name and address, identification number, results of lab tests, and physicians' notes. Images include Xrays and scans from computer tomography, magnetic resonance imaging, and ultrasound. Examples of physiological signals are the electrocardiogram (ECG), the electroencephalogram (EEG), and blood pressure tracings.

Quite different systems are necessary to manipulate each of these three types of data. Alphanumeric data are generally managed and organized into a database using a general-purpose mainframe computer.

Image data are traditionally archived on film. However, we are evolving toward picture archiving and communication systems (PACS) that will store images in digitized form on optical disks and distribute them on demand over a high-speed local area network (LAN) to very high resolution graphics display monitors located throughout a hospital.

On the other hand, physiological signals like those that are monitored during surgery in the operating room require real-time processing. The clinician must know immediately if the instrument finds abnormal readings as it analyzes the continuous data.

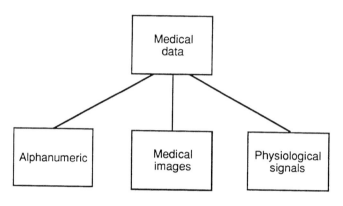

Figure 1.1 Types of medical data.

It is this final type of data on which we concentrate in this book. One of the most monitored signals is the ECG, so we use it as the example signal to process in many examples.

1.2 WHAT IS A MEDICAL INSTRUMENT?

There are many different types of medical instruments. The ones on which we concentrate in this book are those that monitor and analyze physiological signals from a patient. Figure 1.2 shows a block diagram that characterizes such instruments. Sensors measure the patient's physiological signals and produce electrical signals (generally time-varying voltages) that are analogs of the actual signals.

A set of electrodes may be used to sense a potential difference on the body surface such as an ECG or EEG. Sensors of different types are available to transduce into voltages such variables as body core temperature and arterial blood pressure. The electrical signals produced by the sensors interface to a processor which is responsible for processing and analysis of the signals. The processor block typically includes a microprocessor for performing the necessary tasks. Many instruments have the ability to display, record, or distribute through a network either the raw signal captured by the processor or the results of its analysis. In some instruments, the processor performs a control function. Based on the results of signal analysis, the processor might instruct a controller to do direct therapeutic intervention on a patient (closed loop control) or it may signal a person that there is a problem that requires possible human intervention (open loop control).

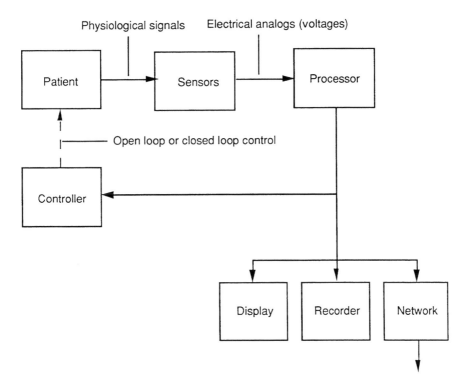

Figure 1.2 Basic elements of a medical instrumentation system.

Let us consider two types of medical instrumentation and see how they fit this block diagram. The first is an intensive care unit (ICU) system, a large set of instrumentation that monitors a number of patients simultaneously. The second is a cardiac pacemaker so small that it must fit inside the patient.

In the case of the ICU, there are normally several sensors connected to each patient receiving intensive care, and the processor (actually usually more than one processor) monitors and analyzes all of them. If the processor discovers an abnormality, it alerts the medical staff, usually with audible alarms. A display permits the staff to see raw data such as the ECG signals for each patient and also data obtained from the analysis such as numerical readouts of heart rate and blood pressure. The network connects the bedside portion of the instrumentation to a central console in the ICU. Another network might connect the ICU system to other databases remotely located in the hospital. An example of a closed loop device that is sometimes used is an infusion pump. Sensors monitor fluid loss as the amount of urine collected from the patient, then the processor instructs the pump to infuse the proper amount of fluid into the patient to maintain fluid balance, thereby acting as a therapeutic device.

Now consider Figure 1.2 for the case of the implanted cardiac pacemaker. The sensors are electrodes mounted on a catheter that is placed inside the heart. The processor is usually a specialized integrated circuit designed specifically for this ultra-low-power application rather than a general-purpose microprocessor. The processor monitors the electrogram from the heart and analyzes it to determine if the heart is beating by itself. If it sees that the heart goes too long without its own stimulus signal, it fires an electrical stimulator (the controller in this case) to inject a large enough current through the same electrodes as those used for monitoring. This stimulus causes the heart to beat. Thus this device operates as a closed loop therapy delivery system. The early pacemakers operated in an open loop fashion, simply driving the heart at some fixed rate regardless of whether or not it was able to beat in a normal physiological pattern most of the time. These devices were far less satisfactory than their modern *intelligent* cousins. Normally a microprocessor-based device outside the body placed over a pacemaker can communicate with it through telemetry and then display and record its operating parameters. Such a device can also set new operating parameters such as amplitude of current stimulus. There are even versions of such devices that can communicate with a central clinic over the telephone network.

Thus, we see that the block diagram of a medical instrumentation system serves to characterize many medical care devices or systems.

1.3 ITERATIVE DEFINITION OF MEDICINE

Figure 1.3 is a block diagram that illustrates the operation of the medical care system. Data collection is the starting point in health care. The clinician asks the patient questions about medical history, records the ECG, and does blood tests and other tests in order to define the patient's problem. Of course medical instruments help in some aspects of this data collection process and even do some preprocessing of the data. Ultimately, the clinician analyzes the data collected and decides what is the basis of the patient's problem. This decision or diagnosis leads the clinician to prescribe a therapy. Once the therapy is administered to the patient, the process continues around the closed loop in the figure with more data collection and analysis until the patient's problem is gone.

The function of the medical instrument of Figure 1.2 thus appears to be a model of the medical care system itself.

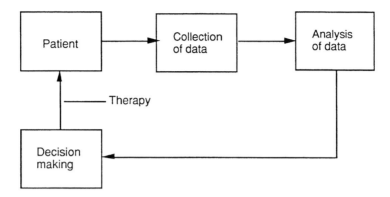

Figure 1.3 Basic elements of a medical care system.

1.4 EVOLUTION OF MICROPROCESSOR-BASED SYSTEMS

In the last decade, the microcomputer has made a significant impact on the design of biomedical instrumentation. The natural evolution of the microcomputer-based instrument is toward more *intelligent* devices. More and more computing power and memory are being squeezed into smaller and smaller spaces. The commercialization of laptop PCs with significant computing power has accelerated the technology of the battery-powered, patient-worn portable instrument. Such an instrument can be truly a *personal* computer looking for problems specific to a given patient during the patient's daily routines. The ubiquitous PC itself evolved from minicomputers that were developed for the biomedical instrumentation laboratory, and the PC has become a powerful tool in biomedical computing applications. As we look to the future, we see the possibility of developing instruments to address problems that could not be previously approached because of considerations of size, cost, or power consumption.

The evolution of the microcomputer-based medical instrument has followed the evolution of the microprocessor itself (Tompkins and Webster, 1981). Figure 1.4 shows a plot of the number of transistors in Intel microprocessors as a function of time. The microprocessor is now more than 20 years old. It has evolved from modest beginnings as an integrated circuit with 2,000 transistors (Intel 4004) in 1971 to the powerful central processing units of today having more than 1,000,000 transistors (e.g., Intel i486 and Motorola 68040). One of the founders of Intel named Moore observed that the number of functional transistors that can be put on a single piece of silicon doubles about every two years. The solid line in the figure represents this observation, which is now known as Moore's Law. The figure shows that Intel's introduction of microprocessors, to date, has followed Moore's Law exceptionally well. The company has predicted that they will be able to continue producing microprocessors with this exponential growth in the number of transistors per microprocessor until at least the end of this century. Thus, in less than a decade, the microprocessor promises to become superpowerful as a parallel processing device with 100 million transistors on one piece of silicon. It most likely will be more powerful than any of today's supercomputers, will certainly be part of a desktop computer, and possibly will be powerable by batteries so that it can be used in portable devices.

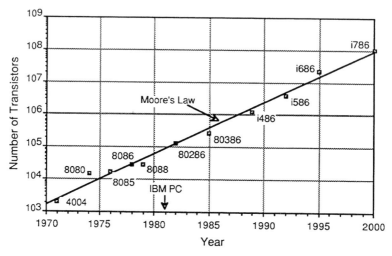

Figure 1.4 The evolution of the microprocessor. The number of transistors in a microprocessor has increased exponentially throughout the history of the device. The trend is expected to continue into the future.

The evolution of the microprocessor from its early beginnings in 1971 as a primitive central processing unit to the powerful component of today has made a significant impact on the design of biomedical instrumentation. More computing power and memory are being squeezed into fewer integrated circuits to provide increasingly more powerful instruments. The PC itself has become a powerful tool in biomedical computing applications. In the future, we will be able to develop new medical instruments to address problems that were previously not solvable. This possibility exists because microprocessor-based systems continuously increase in computing power and memory while decreasing in size, cost, and power consumption.

1.4.1 Evolution of the personal computer

Figure 1.5 shows the history of the development of the computer from the first mechanical computers such as those built by Charles Babbidge in the 1800s to the modern personal computers, the IBM PC and the Apple Macintosh. The only computers prior to the twentieth century were mechanical, based on gears and mechanical linkages.

In 1941 a researcher named Atanasoff demonstrated the first example of an electronic digital computer. This device was primitive even compared to today's four-function pocket calculator. The first serious digital computer called ENIAC (Electronic Numerical Integrator And Calculator) was developed in 1946 at the Moore School of Electrical Engineering of the University of Pennsylvania. Still simple compared to the modern PC, this device occupied most of the basement of the Moore School and required a substantial air conditioning system to cool the thousands of vacuum tubes in its electronic brain.

The invention of the transistor led to the Univac I, the first commercial computer. Several other companies including IBM subsequently put transistorized computers into the marketplace. In 1961, researchers working at Massachusetts Institute of Technology and Lincoln Labs used the technology of the time to build a novel minicomputer quite unlike the commercial machines. This discrete-component, transistorized minicomputer with magnetic core memory called the LINC (Laboratory INstrument Computer) was the most significant historical development in the evolution of the PC.

The basic design goal was to transform a general-purpose computer into a laboratory instrument for biomedical computing applications. Such a computer, as its designers envisioned, would have tremendous versatility because its function as an instrument could be completely revised simply by changing the program stored in its memory. Thus this computer would perform not only in the classical computer sense as an equation solving device, but also by reprogramming (software), it would be able to mimic many other laboratory instruments.

The LINC was the most successful minicomputer used for biomedical applications. In addition, its design included features that we have come to expect in modern PCs. In particular, it was the world's first interactive computer. Instead of using punched cards like the other computers of the day, the LINC had a keyboard and a display so that the user could sit down and program it directly. This was the first digital computer that had an interactive graphics display and that incorporated knobs that were functionally equivalent to the modern joystick. It also had built-in signal conversion and instrument interfacing hardware, with a compact, reliable digital tape recorder, and with sound generation capability. You could capture an ECG directly from a patient and show the waveform on the graphics display.

The LINC would have been the first personal computer if it had been smaller (it was about the size of a large refrigerator) and less expensive (it cost about $50,000 in kit form). It was the first game computer. Programmers wrote software for a two-player game called Spacewar. Each player controlled the velocity and direction of a spaceship by turning two knobs. Raising a switch fired a missile at the opposing ship. There were many other games such as pong and music that included an organ part from Bach as well as popular tunes.

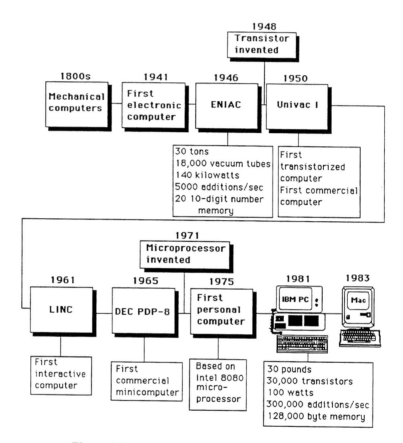

Figure 1.5 The evolution of the personal computer.

The LINC was followed by the world's first commercial minicomputer, which was also made of discrete components, the Digital Equipment Corporation PDP-8. Subsequently Digital made a commercial version of the LINC by combining the LINC architecture with the PDP-8 to make a LINC-8. Digital later introduced a more modern version of the LINC-8 called the PDP-12. These computers were phased out of Digital's product line some time in the early 1970s. I have a special fondness for the LINC machines since a LINC-8 was the first computer that I programmed that was interactive, could display graphics, and did not require the use of awkward media like punched cards or punched paper tape to program it. One of the programs that I wrote on the LINC-8 in the late 1960s computed and displayed the vectorcardiogram loops of patients (see Chapter 2). Such a program is easy to implement today on the modern PC using a high-level computing language such as Pascal or C.

Although invented in 1971, the first microprocessors were poor central processing units and were relatively expensive. It was not until the mid-1970s when useful 8-bit microprocessors such as the Intel 8080 were readily available. The first advertised microcomputer for the home appeared on the cover of *Popular Electronics Magazine* in January 1975. Called the Altair 8800, it was based on the Intel 8080 microprocessor and could be purchased as a kit. The end of the decade was full of experimentation and new product development leading to the introduction of PCs like the Apple II and microcomputers from many other companies.

1.4.2 The ubiquitous PC

A significant historical landmark was the introduction of the IBM PC in 1981. On the strength of its name alone, IBM standardized the personal desktop computer. Prior to the IBM PC, the most popular computers used the 8-bit Zilog Z80 microprocessor (an enhancement of the Intel 8080) with an operating system called CP/M (Control Program for Microprocessors). There was no standard way to format a floppy disk, so it was difficult to transfer data from one company's PC to another. IBM singlehandedly standardized the world almost overnight on the 16-bit Intel 8088 microprocessor, Microsoft DOS (Disk Operating System), and a uniform floppy disk format that could be used to carry data from machine to machine. They also stimulated worldwide production of IBM PC compatibles by many international companies. This provided a standard computing platform for which software developers could write programs. Since so many similar computers were built, inexpensive, powerful application programs became plentiful. This contrasted to the minicomputer marketplace where there are relatively few similar computers in the field, so a typical program is very expensive and the evolution of the software is relatively slow.

Figure 1.6 shows how the evolution of the microprocessor has improved the performance of desktop computers. The figure is based on the idea that a complete PC at any given time can be purchased for about $5,000. For this cost, the number of MIPS (million instructions per second) increases with every new model because of the increasing power of the microprocessor. The first IBM PC introduced in 1981 used the Intel 8088 microprocessor, which provided 0.1 MIPS. Thus, it required 10 PCs at $5,000 each or $50,000 to provide one MIPS of computational power. With the introduction of the IBM PC/AT based on the Intel 80286 microprocessor, a single $5,000 desktop computer provided one MIPS. The most recent IBM PS computers using the Intel i486 microprocessor deliver 10 MIPS for that same $5,000. A basic observation is that the cost per MIPS of computing power is decreasing logarithmically in desktop computers.

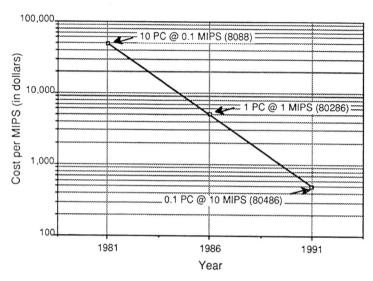

Figure 1.6 The inverse relationship between computing power and cost for desktop computers. The cost per MIPS (million instructions per second) has decreased logarithmically since the introduction of the IBM PC.

Another important PC landmark was the introduction of the Apple Macintosh in 1983. This computer popularized a simple, intuitive user-to-machine interface. Since that time, there have been a number of attempts to implement similar types of graphical user interface (GUI) for the IBM PC platform, and only recently have practical solutions come close to reality.

More than a decade has elapsed since the introduction of the IBM PC, and most of the changes in the industry have been evolutionary. Desktop PCs have continued to evolve and improve with the evolution of the technology, particularly the microprocessor itself. We now have laptop and even palmtop PC compatibles that are portable and battery powered. We can carry around a significant amount of computing power wherever we go.

Figure 1.7 shows how the number of electronic components in a fully functional PC has decreased logarithmically with time and will continue to decrease in the future. In 1983, about 300 integrated circuits were required in each PC, of which about half were the microprocessor and support logic and the other half made up the 256-kbyte memory. Today half of the parts are still dedicated to each function, but a complete PC can be built with about 18 ICs. In the past six years, the chip count in a PC has gone from about 300 integrated circuits in a PC with a 256-kbyte memory to 18 ICs in a modern 2-Mbyte PC. By the mid-1990s, it is likely that a PC with 4-Mbyte memory will be built from three electronic components, a single IC for all the central processing, a read-only memory (ROM) for the basic input/output software (BIOS), and a 4-Mbyte dynamic random access memory (DRAM) chip for the user's program and data storage. Thus the continuing trend is toward more powerful PCs with more memory in smaller space for lower cost and lower power consumption.

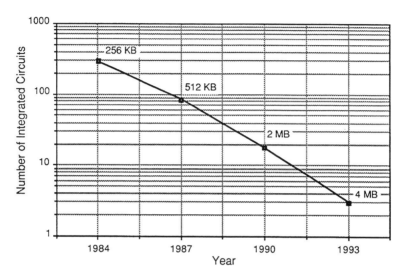

Figure 1.7 The number of components in a PC continues to decrease while the computing performance increases.

In the 1970s, the principal microprocessors used in desktop computers as well as other systems including medical instruments were 8-bit microprocessors. The 1980s were dominated by the 16-bit microprocessors. The 1990s were launched with the 32-bit processor, but the technology's exponential growth will likely lead to useful new architectures on single ICs, such as parallel processors and artificial neural networks.

Figure 1.8 compares the modern PC with the human brain. The figure provides information that gives us insight into the relative strengths and weaknesses of the computer compared to the brain. From the ratios provided, we can clearly see that the personal computer is one or more orders of magnitude heavier, larger, and more power consuming than the human brain. The PC has several orders of magnitude fewer functional computing elements and memory cells in its "brain" than does the human brain.

COMPARISON OF PERFORMANCE OF IBM PC AND HUMAN BRAIN

System	Weight (lbs)	Size (ft^3)	Power (watts)	CPU elements	Memory (bits)	Conduction rate (impulses/s)	Benchmark (additions/s)
IBM PC	30	5	200	10^6 transistors (or equiv.)	10^7	10^5	10^6
Brain	3	0.05	10	10^{10} neurons	10^{20}	10^2	1
Ratio (PC/brain)	10	100	20	10^{-4}	10^{-13}	10^3	10^6

Figure 1.8 The PC and the human brain each have their own characteristic strengths and weaknesses.

Then what good is it? The answer lies in the last two columns of the figure. The speed at which impulses are conducted in the computer so far exceeds the speed of movement of information within the brain that the PC has a very large computation speed advantage over the brain. This is illustrated by a benchmark which asks a human to add one plus one plus one and so on, reporting the sum after each addition. The PC can do this computational task about one million times faster than the human. If the PC is applied to tasks that exploit this advantage, it can significantly outperform the human.

Figure 1.9(a) is an array of numbers, the basic format in which all data must be placed before it can be manipulated by a computer. These numbers are equally spaced amplitude values for the electrocardiogram (ECG) of Figure 1.9(b). They were obtained by sampling the ECG at a rate of 200 samples per second with an analog-to-digital converter. This numerical representation of the ECG is called a digital signal. The human eye-brain system, after years of experience in learning the characteristics of these signals, is particularly good at analyzing the analog waveform itself and deducing whether such a signal is normal or abnormal.

On the other hand, the computer must analyze the array of numbers using software algorithms in order to make deductions about the signal. These algorithms typically include digital signal processing, which is the emphasis of this book, together with decision logic in order to analyze biomedical signals as well as medical images. It is the enormous speed of the computer at manipulating numbers that makes many such algorithms possible.

0 0 0 0 2 5 8 10 13 14 14 14 14 12 11 9 7 5 4 2 1 1 0 0 1 1 1 1 1 2 2 2 3 3 3 3 3 3 3 3
3 3 6 11 20 33 51 72 91 103 105 96 77 53 27 5 -11 -23 -28 -28 -23 -17 -10 -5 -1
0 1 2 1 1 1 1 0 0 0 0 0 0 0 0 0 0 0 1 1 2 2 3 3 4 4 5 6 7 8 8 9 10 10 11 11 12 12
12 13 14 16 18 20 22 24 27 29 31 34 37 39 42 44 47 49 52 54 55 56 57 57 58 58
57 57 56 56 54 52 50 47 43 40 36 33 29 26 23 20 17 14 12 10 8 7 5 3 2 1 1 0 0 0

(a)

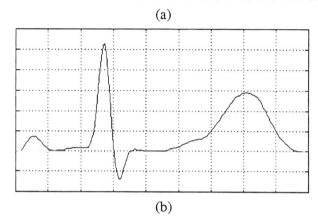

(b)

Figure 1.9 Two views of an electrocardiogram. (a) The computer view is an array of numbers that represent amplitude values as a function of time. (b) The human view is a time-varying waveform.

1.5 THE MICROCOMPUTER-BASED MEDICAL INSTRUMENT

The progress in desktop and portable computing in the past decade has provided the means with the PC or customized microcomputer-based instrumentation to develop solutions to biomedical problems that could not be approached before. One of our personal interests has been the design of portable instruments that are light, compact, and battery powered (Tompkins, 1981). A typical instrument of this type is truly a personal computer since it is programmed to monitor signals from transducers or electrodes mounted on the person who is carrying it around.

1.5.1 Portable microcomputer-based instruments

One example of a portable device is the portable arrhythmia monitor which monitors a patient's electrocardiogram from chest electrodes and analyzes it in real time to determine if there are any heart rhythm abnormalities. We designed a prototype of such a device more than a decade ago (Tompkins, 1978). Because of the technology available at that time, this device was primitive compared with modern commercially available portable arrhythmia monitors. The evolution of the technology also permits us to think of even more extensions that we can make. Instead of just assigning a heart monitoring device to follow a patient after discharge from the hospital, we can now think of designing a device that would help diagnose the heart abnormality when the patient arrives in the emergency room. With a careful design, the same device might go with the patient to monitor the cardiac problem during surgery in the operating room, continuously learning the unique character- istics of that patient's heart rhythms. The device could follow the patient through- out the hospital stay, alerting the hospital staff to possible problems in the intensive care unit, in the regular hospital room, and even in the hallways as the patient walks to the cafeteria. The device could then accompany the patient home, provid- ing continuous monitoring that is not now practical to do, during the critical times following open heart surgery (Tompkins, 1988). Chapter 13 discusses the concept of a portable arrhythmia monitor in greater detail.

There are many other examples of portable biomedical instruments in the mar- ketplace and in the research lab. One other microcomputer-based device that we contributed to developing is a calculator-size product called the CALTRAC that uses a miniature accelerometer to monitor the motion of the body. It then converts this activity measurement to the equivalent number of calories and displays the cumulative result on an LCD display (Doumas et al., 1982). There is now an im- planted pacemaker that uses an accelerometer to measure the level of a patient's activity in order to adjust the pacing rate.

We have also developed a portable device that monitors several pressure chan- nels from transducers on a catheter placed in the esophagus. It analyzes the signals for pressure changes characteristic of swallowing, then records these signals in its semiconductor memory for later transfer to an IBM PC where the data are further analyzed (Pfister et al., 1989).

Another portable device that we designed monitors pressure sensors placed in the shoes to determine the dynamic changes in pressure distribution under the foot for patients such as diabetics who have insensate feet (Mehta et al., 1989).

1.5.2 PC-based medical instruments

The economy of mass production has led to the use of the desktop PC as the central computer for many types of biomedical applications. Many companies use PCs for such applications as sampling and analyzing physiological signals, maintaining equipment databases in the clinical engineering department of hospitals, and simulation and modeling of physiological systems.

You can configure the PC to have user-friendly, interactive characteristics much like the LINC. This is an important aspect of computing in the biomedical laboratory. The difference is that the PC is a much more powerful computer in a smaller, less expensive box. Compared to the LINC of two decades ago, the PC has more than 100 times the computing power and 100 times the memory capacity in one-tenth the space for one-tenth the cost. However, the LINC gave us tremendous insight into what the PC should be like long before it was possible to build a personal computer.

We use the PC as a general-purpose laboratory tool to facilitate research on many biomedical computing problems. We can program it to execute an infinite variety of programs and adapt it for many applications by using custom hardware interfaces. For example, the PC is useful in rehabilitation engineering. We have designed a system for a blind person that converts visible images to tactile (touch) images. The PC captures an image from a television camera and stores it in its memory. A program presents the image piece by piece to the blind person's fingertip by activating an array of tactors (i.e., devices that stimulate the sense of touch) that are pressed against his/her fingertip. In this way, we use the PC to study the ability of a blind person to "see" images with the sense of touch (Kaczmarek et al., 1985; Frisken-Gibson et al., 1987).

One of the applications that we developed based on an Apple Macintosh II computer is electrical impedance tomography—EIT (Yorkey et al., 1987; Woo et al., 1989; Hua et al., 1991). Instead of the destructive radiation used for the familiar computerized tomography techniques, we inject harmless high-frequency currents into the body through electrodes and measure the resistances to the flow of electricity at numerous electrode sites. This idea is based on the fact that body organs differ in the amount of resistance that they offer to electricity. This technology attempts to image the internal organs of the human body by measuring impedance through electrodes placed on the body surface.

The computer controls a custom-built 32-channel current generator that injects patterns of high-frequency (50-kHz) currents into the body. The computer then samples the body surface voltage distribution resulting from these currents through an analog-to-digital converter. Using a finite element resistivity model of the thorax and the boundary measurements, the computer then iteratively calculates the resistivity profile that best satisfies the measured data. Using the standard graphics display capability of the computer, an image is then generated of the transverse body section resistivity. Since the lungs are high resistance compared to the heart and other body tissues, the resistivity image provides a depiction of the organ system in the body. In this project the Macintosh does all the instrumentation tasks including control of the injected currents, measurement of the resistivities, solving the computing-intensive algorithms, and presenting the graphical display of the final image.

There are many possible uses of PCs in medical instrumentation (Tompkins, 1986). We have used the IBM PC to develop signal processing and artificial neural network (ANN) algorithms for analysis of the electrocardiogram (Pan and Tompkins, 1985; Hamilton and Tompkins, 1986; Xue et al., 1992). These studies have also included development of techniques for data compression to reduce the amount of storage space required to save ECGs (Hamilton and Tompkins, 1991a, 1991b).

1.6 SOFTWARE DESIGN OF DIGITAL FILTERS

In addition to choosing a personal computer hardware system for laboratory use, we must make additional software choices. The types of choices are frequently closely related and limited by the set of options available for a specific hardware system. Figure 1.10 shows that there are three levels of software between the hardware and the real-world environment: the operating system, the support software, and the application software (the shaded layer). It is the application software that makes the computer behave as a medical instrument. Choices of software at all levels significantly influence the kinds of applications that a system can address.

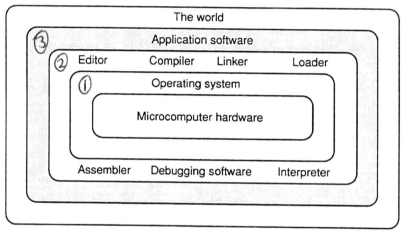

Figure 1.10 Three levels of software separate a hardware microcomputer system from the real-world environment. They are the operating system, the support software, and the application software.

Two major software selections to be made are (1) choice of the disk operating system (DOS) to support the development task, and (2) choice of the language to implement the application. Although many different combinations of operating system and language are able to address the same types of applications, these choices frequently are critical since certain selections are clearly better than others for some types of applications. Of course these two choices are influenced significantly by the initial hardware selection, by personal biases, and by the user's level of expertise.

1.6.1 Disk operating systems

Our applications frequently involve software implementation of real-time signal processing algorithms, so we orient the discussions around this area. Real-time means different things to different people in computing. For our applications, consider real-time computing to be what is required of video arcade game machines. The microcomputer that serves as the central processing unit of the game machine must do all its computing and produce its results in a time frame that appears to the user to be instantaneous. The game would be far less fun if, each time you fired a missile, the processor required a minute or two to determine the missile's trajectory and establish whether or not it had collided with an enemy spacecraft.

A typical example of the need for real-time processing in biomedical computing is in the analysis of electrocardiograms in the intensive care unit of the hospital. In the typical television medical drama, the ailing patient is connected to a monitor that beeps every time the heart beats. If the monitor's microcomputer required a minute or two to do the complex pattern recognition required to recognize each valid heartbeat and then beeped a minute or so after the actual event, the device would be useless. The challenge in real-time computing is to develop programs to implement procedures (algorithms) that appear to occur instantaneously (actually a given task may take several milliseconds).

One DOS criterion to consider in the real-time environment is the compromise between flexibility and usability. Figure 1.11 is a plot illustrating this compromise for several general-purpose microcomputer DOSs that are potentially useful in developing solutions to many types of problems including real-time applications. As the axes are labeled, the most user-friendly, flexible DOS possible would plot at the origin. Any DOS with an optimal compromise between usability and flexibility would plot on the 45-degree line.

A DOS like Unix has a position on the left side of the graph because it is very flexible, thereby permitting the user to do any task characteristic of an operating system. That is, it provides the capability to maximally manipulate a hardware/software system with excellent control of input/output and other facilities. It also provides for multiple simultaneous users to do multiple simultaneous tasks (i.e., it is a multiuser, multitasking operating system). Because of this great flexibility, Unix requires considerable expertise to use all of its capabilities. Therefore it plots high on the graph.

On the other hand, the Macintosh is a hardware/software DOS designed for ease of use and for graphics-oriented applications. Developers of the Macintosh implemented the best user-to-machine interface that they could conceive of by sacrificing a great deal of the direct user control of the hardware system. The concept was to produce a personal computer that would be optimal for running application programs, not a computer to be used for writing new application programs. In fact Apple intended that the Lisa would be the development system for creating new Macintosh programs.

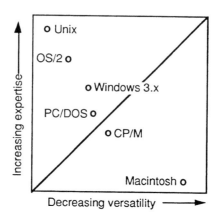

Figure 1.11 Disk operating systems–the compromise between DOS versatility and user expertise in real-time applications.

Without training, an individual can sit down and quickly learn to use a Macintosh because its operation is, by design, intuitive. The Macintosh DOS plots low and to the right because it is very user-friendly but cannot necessarily be used by the owner to solve any generalized problem. In fact, the Macintosh was the first personal computer in history that was sold without any language as part of the initial package. When the Macintosh was first introduced, no language was available for it, not even the ubiquitous BASIC that came free with almost every other computer at the time, including the original IBM PC.

The 8-bit operating system, CP/M (Control Program/Microprocessors), was a popular operating systems because it fell near the compromise line and represented a reasonable mixture of ease of use and flexibility. Also it could be implemented with limited memory and disk storage capability. CP/M was the most popular operating system on personal computers based on 8-bit microprocessors such as the Zilog Z80. PC DOS (or the generic MS DOS) was modeled after CP/M to fall near the compromise line. It became the most-used operating system on 16-bit personal computers, such as the IBM PC and its clones, that are based on the Intel 8086/8088 microprocessor or other 80x86 family members.

On the other hand, Unix is not popular on PCs because it requires a great deal of memory and hard disk storage to achieve its versatility. The latest Unix look-alike operating systems for the IBM PC and the Macintosh typically require significant memory and many megabytes of hard disk storage. This is compared to CP/M on an 8-bit system that normally was implemented in less than 20 kbytes of storage space and PC/DOS on an IBM PC that requires less than 100 kbytes.

At this writing, Unix is the workstation operating system of choice. For many applications, it may end up to be the DOS of choice. Indeed, Unix or a close clone of it may ultimately provide the most accepted answer to the problem of linking PCs together through a local area network (LAN). By its very design, Unix provides multitasking, a feature necessary for LAN implementation. Incidentally, the fact that Unix is written in the C language gives it extraordinary transportability, facilitating its implementation on computers ranging from PCs to supercomputers.

For real-time digital filtering applications, Unix is not desirable because of its overhead compared to PC/DOS. In order to simultaneously serve multiple tasks, it must use up some computational speed. For the typical real-time problem, there is no significant speed to spare for this overhead. Real-time digital signal processing requires a single-user operating system. You must be able to extract the maximal performance from the computer and be able to manipulate its lowest level resources such as the hardware interrupt structure.

The trends for the future will be toward Macintosh-type operating systems such as Windows and OS/2. These user-friendly systems sacrifice a good part of the generalized computing power to the human-to-machine interface. Each DOS will be optimized for its intended application area and will be useful primarily for that area. Fully implemented versions of OS/2 will most likely require such large portions of the computing resources that they will have similar liabilities to those of Unix in the real-time digital signal processing environment.

Unfortunately there are no popular operating systems available that are specifically designed for such real-time applications as digital signal processing. A typical DOS is designed to serve the largest possible user base; that is, to be as general purpose as possible. In an IBM PC, the DOS is mated to firmware in the ROM BIOS (Basic Input/Output System) to provide a general, orderly way to access the system hardware. Use of high-level language calls to the BIOS to do a task such as display of graphics reduces software development time because assembly language is not required to deal directly with the specific integrated circuits that control the graphics. A program developed with high-level BIOS calls can also be easily transported to other computers with similar resources. However, the BIOS firmware is general purpose and has some inefficiencies. For example, to improve the speed of graphics refresh of the screen, you can use assembly language to bypass the BIOS and write directly to the display memory. However this speed comes at the cost of added development time and loss of transportability.

Of course, computers like the NEXT computer are attempting to address some of these issues. For example, the NEXT has a special shell for Unix designed to make it more user-friendly. It also includes a built-in digital signal processing (DSP) chip to facilitate implementation of signal processing applications.

1.6.2 Languages

Figure 1.12 shows a plot of the time required to write an application program as a function of run-time speed. This again is plotted for the case of real-time applications such as digital signal processing. The best language for this application area would plot at the origin since this point represents a program with the greatest run-time speed and the shortest development time. The diagonal line maps the best compromise language in terms of the run-time speed compared to the software design time necessary to implement an application. Of course there are other considerations for choosing a language, such as development cost and size of memory space available in an instrument. Also most of the languages plotted will not, by themselves, solve the majority of real-time problems, especially of the signal processing type.

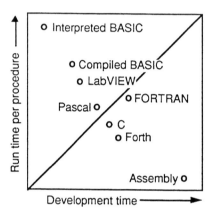

Figure 1.12 Languages–the compromise between development time and run-time speed in real-time applications.

The time to complete a procedure at run time is generally quite long for interpreted languages (that is, they are computationally slow). These interpreted languages like interpreted BASIC (Beginner's All-Purpose Symbolic Instruction Code) are generally not useful for real-time instrumentation applications, and they plot a great distance from the diagonal line.

The assembly language of a microprocessor is the language that can extract the greatest run-time performance because it provides for direct manipulation of the architecture of the processor. However it is also the most difficult language for writing programs, so it plots far from the optimal language line. Frequently, we must resort to this language in high-performance applications.

Other high-level languages such as FORTRAN and Pascal plot near the diagonal indicating that they are good compromises in terms of the trade-off between program development time and run time per procedure but do not usually produce code with enough run-time speed for real-time signal processing applications. Many applications are currently implemented by combining one of these languages with assembly language routines. FORTRAN was developed for solving equations (i.e., **FOR**mula **TRAN**slation) and Pascal was designed for teaching students structured programming techniques.

After several years of experience with a new microprocessor, software development companies are able to produce enhanced products. For example, modern versions of Pascal compilers developed for PCs have a much higher performance-to-price ratio than any Pascal compiler produced more than a decade ago.

Forth is useful for real-time applications, but it is a nontraditional, stack-oriented language so different from other programming languages that it takes some time for a person to become a skilled programmer. Documentation of programs is also difficult due to the flexibility of the language. Thus, a program developed in Forth typically is a one-person program. However, there are several small versions of the Forth compiler built into the same chip with a microprocessor. These implementations promote its use particularly for controller applications.

LabVIEW (National Instruments) is a visual computing language available only for the Macintosh that is optimized for laboratory applications. Programming is accomplished by interconnecting functional blocks (i.e., icons) that represent processes such as Fourier spectrum analysis or instrument simulators (i.e., virtual instruments). Thus, unlike traditional programming achieved by typing command statements, LabVIEW programming is purely graphical, a block diagram language. Although it is a relatively fast compiled language, LabVIEW is not optimized for real-time applications; its strengths lie particularly in the ability to acquire and process data in the laboratory environment.

 The C language, which is used to develop modern versions of the Unix operating system, provides a significant improvement over assembly language for implementing most applications (Kernighan and Ritchie, 1978). It is the current language of choice for real-time programming. It is an excellent compromise between a low-level assembly language and a high-level language. C is standardized and structured. There are now several versions of commercial C++ compilers available for producing object-oriented software.

C programs are based on functions that can be evolved independently of one another and put together to implement an application. These functions are to software what black boxes are to hardware. If their I/O properties are carefully specified in advance, functions can be developed by many different software designers working on different aspects of the same project. These functions can then be linked together to implement the software design of a system.

Most important of all, C programs are transportable. By design, a program developed in C on one type of processor can be relatively easily transported to another. Embedded machine-specific functions such as those written in assembly language can be separated out and rewritten in the native code of a new architecture to which the program has been transported.

1.7 A LOOK TO THE FUTURE

As the microprocessor and its parent semiconductor technologies continue to evolve, the resulting devices will stimulate the development of many new types of medical instruments. We cannot even conceive of some of the possible applications now, because we cannot easily accept and start designing for the significant advances that will be made in computing in the next decade. With the 100-million-transistor microprocessor will come personal supercomputing. Only futurists can contemplate ways that we individually will be able to exploit such computing power. Even the nature of the microprocessor as we now know it might change more toward the architecture of the artificial neural network, which would lead to a whole new set of pattern recognition applications that may be more readily solvable than with today's microprocessors.

The choices of a laboratory computer, an operating system, and a language for a task must be done carefully. The IBM-compatible PC has emerged as a clear computer choice because of its widespread acceptance in the marketplace. The fact that so many PCs have been sold has produced many choices of hardware add-ons developed by numerous companies and also a wide diversity of software application programs and compilers. By default, IBM produced not only a hardware standard but also the clear-cut choice of the PC DOS operating system for the first decade of the use of this hardware. Although there are other choices now, DOS is still alive. Many will choose to continue using DOS for some time to come, adding to it a graphical user interface (GUI) such as that provided by Windows (Microsoft).

This leaves only the choice of a suitable language for your application area. My choice for biomedical instrumentation applications is C. In my view this is a clearly superior language for real-time computing, for instrumentation software design, and for other biomedical computing applications.

The hardware/software flexibility of the PC is permitting us to do research in areas that were previously too difficult, too expensive, or simply impossible. We have come a long way in biomedical computing since those innovators put together that first PC-like LINC almost three decades ago. Expect the PC and its descendants to stimulate truly amazing accomplishments in biomedical research in the next decade.

1.8 REFERENCES

Doumas, T. A., Tompkins, W. J., and Webster, J. G. 1982. An automatic calorie measuring device. *IEEE Frontiers of Eng. in Health Care*, **4**: 149–51.

Frisken-Gibson, S., Bach-y-Rita, P., Tompkins, W. J., and Webster, J. G. 1987. A 64-solenoid, 4-level fingertip search display for the blind. *IEEE Trans. Biomed. Eng.*, **BME-34**(12): 963–65.

Hamilton, P. S., and Tompkins, W. J. 1986. Quantitative investigation of QRS detection rules using the MIT/BIH arrhythmia database. *IEEE Trans. Biomed. Eng.*, **BME-33**(12): 1157–65.

Hua, P., Woo, E. J., Webster, J. G., and Tompkins, W. J. 1991. Iterative reconstruction methods using regularization and optimal current patterns in electrical impedance tomography. *IEEE Trans. Medical Imaging*, **10**(4): 621–28.

Kaczmarek, K., Bach-y-Rita, P., Tompkins, W. J., and Webster, J. G. 1985. A tactile vision substitution system for the blind: computer-controlled partial image sequencing. *IEEE Trans. Biomed. Eng.*, **BME-32**(8):602–08.

Kernighan, B. W., and Ritchie, D. M. 1978. *The C programming language*. Englewood Cliffs, NJ: Prentice Hall.

Mehta, D., Tompkins, W. J., Webster, J. G., and Wertsch, J. J. 1989. Analysis of foot pressure waveforms. *Proc. Annual International Conference of the IEEE Engineering in Medicine and Biology Society*, pp. 1487–88.

Pan, J. and Tompkins, W. J. 1985. A real-time QRS detection algorithm. *IEEE Trans. Biomed. Eng.*, **BME-32**(3): 230–36.

Pfister, C., Harrison, M. A., Hamilton, J. W., Tompkins, W. J., and Webster, J. G. 1989. Development of a 3-channel, 24-h ambulatory esophageal pressure monitor. *IEEE Trans. Biomed. Eng.*, **BME-36**(4): 487–90.

Tompkins, W. J. 1978. A portable microcomputer-based system for biomedical applications. *Biomed. Sci. Instrum.*, **14**: 61–66.

Tompkins, W. J. 1981. Portable microcomputer-based instrumentation. In H. S. Eden and M. Eden (eds.) *Microcomputers in Patient Care*. Park Ridge, NJ: Noyes Medical Publications, pp. 174–81.

Tompkins, W. J. 1985. Digital filter design using interactive graphics on the Macintosh. *Proc. of IEEE EMBS Annual Conf.*, pp. 722–26.

Tompkins, W. J. 1986. Biomedical computing using personal computers. *IEEE Engineering in Medicine and Biology Magazine*, **5**(3): 61–64.

Tompkins, W. J. 1988. Ambulatory monitoring. In J. G. Webster (ed.) *Encyclopedia of Medical Devices and Instrumentation*. New York: John Wiley, **1**:20–28.

Tompkins, W. J. and Webster, J. G. (eds.) 1981. *Design of Microcomputer-based Medical Instrumentation*. Englewood Cliffs, NJ: Prentice Hall.

Woo, E. J., Hua, P., Tompkins, W. J., and Webster, J. G. 1989. 32-electrode electrical impedance tomograph – software design and static images. *Proc. Annual International Conference of the IEEE Engineering in Medicine and Biology Society*, pp. 455–56.

Xue, Q. Z., Hu, Y. H. and Tompkins, W. J. 1992. Neural-network-based adaptive matched filtering for QRS detection. *IEEE Trans. Biomed. Eng.*, **BME-39**(4): 317–29.

Yorkey, T., Webster, J. G., and Tompkins, W. J. 1987. Comparing reconstruction algorithms for electrical impedance tomography. *IEEE Trans. Biomed. Eng.*, **BME-34**(11):843–52.

1.9 STUDY QUESTIONS

1.1 Compare operating systems for support in developing real-time programs. Explain the relative advantages and disadvantages of each for this type of application.

1.2 Explain the differences between interpreted, compiled, and integrated-environment compiled languages. Give examples of each type.

1.3 List two advantages of the C language for real-time instrumentation applications. Explain why they are important.

2

Electrocardiography

Willis J. Tompkins

One of the main techniques for diagnosing heart disease is based on the electrocardiogram (ECG). The electrocardiograph or ECG machine permits deduction of many electrical and mechanical defects of the heart by measuring ECGs, which are potentials measured on the body surface. With an ECG machine, you can determine the heart rate and other cardiac parameters.

2.1 BASIC ELECTROCARDIOGRAPHY

There are three basic techniques used in clinical electrocardiography. The most familiar is the standard clinical electrocardiogram. This is the test done in a physician's office in which 12 different potential differences called ECG leads are recorded from the body surface of a resting patient. A second approach uses another set of body surface potentials as inputs to a three-dimensional vector model of cardiac excitation. This produces a graphical view of the excitation of the heart called the vectorcardiogram (VCG). Finally, for long-term monitoring in the intensive care unit or on ambulatory patients, one or two ECG leads are monitored or recorded to look for life-threatening disturbances in the rhythm of the heartbeat. This approach is called arrhythmia analysis. Thus, the three basic techniques used in electrocardiography are:

1. Standard clinical ECG (12 leads)
2. VCG (3 orthogonal leads)
3. Monitoring ECG (1 or 2 leads)

Figure 2.1 shows the basic objective of electrocardiography. By looking at electrical signals recorded only on the body surface, a completely noninvasive procedure, cardiologists attempt to determine the functional state of the heart. Although the ECG is an electrical signal, changes in the mechanical state of the heart lead to changes in how the electrical excitation spreads over the surface of the heart, thereby changing the body surface ECG. The study of cardiology is based on the recording of the ECGs of thousands of patients over many years and observing the relationships between various waveforms in the signal and different abnormalities. Thus clinical electrocardiography is largely empirical, based mostly on experiential knowledge. A cardiologist learns the meanings of the various parts of the ECG signal from experts who have learned from other experts.

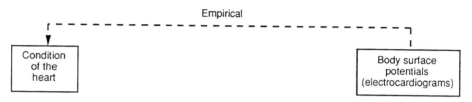

Figure 2.1   The object of electrocardiography is to deduce the electrical and mechanical condition of the heart by making noninvasive body surface potential measurements.

Figure 2.2 shows how the earliest ECGs were recorded by Einthoven at around the turn of the century. Vats of salt water provided the electrical connection to the body. The string galvanometer served as the measurement instrument for recording the ECG.

2.1.1 Electrodes

As time went on, metallic electrodes were developed to electrically connect to the body. An electrolyte, usually composed of salt solution in a gel, forms the electrical interface between the metal electrode and the skin. In the body, currents are produced by movement of ions whereas in a wire, currents are due to the movement of electrons. Electrode systems do the conversion of ionic currents to electron currents.

Conductive metals such as nickel-plated brass are used as ECG electrodes but they have a problem. The two electrodes necessary to acquire an ECG together with the electrolyte and the salt-filled torso act like a battery. A dc offset potential occurs across the electrodes that may be as large or larger than the peak ECG signal. A charge double layer (positive and negative ions separated by a distance) occurs in the electrolyte. Movement of the electrode such as that caused by motion of the patient disturbs this double layer and changes the dc offset. Since this offset potential is amplified about 1,000 times along with the ECG, small changes give rise to large baseline shifts in the output signal. An electrode that behaves in this way is called a polarizable electrode and is only useful for resting patients.

Figure 2.2 Early measurements of an ECG (about 1900) by Einthoven.

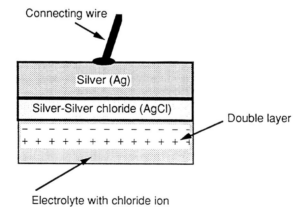

Figure 2.3 A silver-silver chloride ECG electrode. Many modern electrodes have electrolyte layers that are made of a firm gel which has adhesive properties. The firm gel minimizes the disturbance of the charge double layer.

The most-used material for electrodes these days is silver-silver chloride (Ag-AgCl) since it approximates a nonpolarizable electrode. Figure 2.3 shows such an electrode. This type of electrode has a very small offset potential. It has an AgCl layer deposited on an Ag plate. The chloride ions move in the body, in the electrolyte, and in the AgCl layer, where they get converted to electron flow in the Ag plate and in the connecting wire. This approach reduces the dc offset potential to a very small value compared to the peak ECG signal. Thus, movement of the electrode causes a much smaller baseline shift in the amplified ECG than that of a polarizable electrode.

2.1.2 The cardiac equivalent generator

Figure 2.4 shows how a physical model called a cardiac equivalent generator can be used to represent the cardiac electrical activity. The most popular physical model is a dipole current source that is represented mathematically as a time-varying vector which gives rise to the clinical vectorcardiogram (VCG). Einthoven postulated that the cardiac excitation could be modeled as a vector. He also realized that the limbs are like direct connections to points on the torso since the current fluxes set up inside the body by the dipole source flow primarily inside the thorax and do not flow significantly into the limbs. Thus he visualized a situation where electrodes could just as well have been connected to each of the shoulders and to a point near the navel had he not been restricted to using vats of saline.

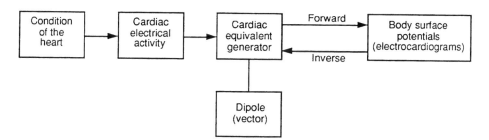

Figure 2.4 Both the electrical and mechanical conditions of the heart are involved in determining the characteristics of the spread of electrical activity over the surface of the heart. A model of this activity is called a cardiac equivalent generator.

Einthoven drew a triangle using as vertices the two shoulders and the navel and observed that the sides of the triangle were about the same length. This triangle, shown in Figure 2.5, has become known as the Einthoven equilateral triangle. If the vector representing the spread of cardiac excitation is known, then the potential difference measured between two limbs (i.e., two vertices of the triangle) is proportional simply to the projection of the vector on the side of the triangle which connects the limbs. The figure shows the relationship between the Einthoven vector and each of the three frontal limb leads (leads I, II, and III). The positive signs show which connection goes to the positive input of the instrumentation amplifier for each lead.

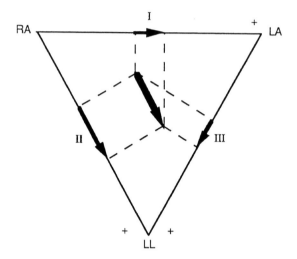

Figure 2.5 Einthoven equilateral triangle. RA and LA are the right and left arms and LL is the left leg.

A current dipole is a current source and a current sink separated by a distance. Since such a dipole has magnitude and direction which change throughout a heartbeat as the cells in the heart depolarize, this leads to the vector representation

$$\mathbf{p}(t) = p_x(t)\,\hat{\mathbf{x}} + p_y(t)\,\hat{\mathbf{y}} + p_z(t)\,\hat{\mathbf{z}} \qquad (2.1)$$

where $\mathbf{p}(t)$ is the time-varying cardiac vector, $p_i(t)$ are the orthogonal components of the vector also called scalar leads, and $\hat{\mathbf{x}}$, $\hat{\mathbf{y}}$, $\hat{\mathbf{z}}$ are unit vectors in the x, y, z directions.

A predominant VCG researcher in the 1950s named Frank shaped a plaster cast of a subject's body like the one shown in Figure 2.6, waterproofed it, and filled it with salt water. He placed a dipole source composed of a set of two electrodes on a stick in the torso model at the location of the heart. A current source supplied current to the electrodes which then produced current fluxes in the volume conductor. From electrodes embedded in the plaster, Frank measured the body surface potential distribution at many thoracic points resulting from the current source. From the measurements in such a study, he found the geometrical transfer coefficients that relate the dipole source to each of the body surface potentials.

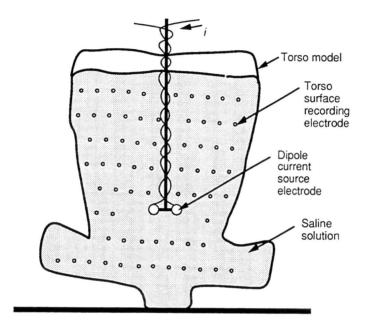

Figure 2.6 Torso model used to develop the Frank lead system for vectorcardiography.

Once the transfer coefficients are known, the forward problem of electrocardiography can be solved for any dipole source. The forward solution provides the potential at any arbitrary point on the body surface for a given cardiac dipole. Expressed mathematically,

$$v_n(t) = t_{nx}\, p_x(t) + t_{ny}\, p_y(t) + t_{nz}\, p_z(t) \tag{2.2}$$

This forward solution shows that the potential $v_n(t)$ (i.e., the ECG) at any point n on the body surface is given by the linear sum of the products of a set of transfer coefficients $[t_{ni}]$ unique to that point and the corresponding orthogonal dipole vector components $[p_i(t)]$. The ECGs are time varying as are the dipole components, while the transfer coefficients are only dependent on the thoracic geometry and inhomogeneities. Thus for a set of k body surface potentials (i.e., leads), there is a set of k equations that can be expressed in matrix form

$$\mathbf{V} = \mathbf{T} \times \mathbf{P} \tag{2.3}$$

where $\mathbf{V}$ is a $k \times 1$ vector representing the time-varying potentials, $\mathbf{T}$ is a $k \times 3$ matrix of transfer coefficients, which are fixed for a given individual, and $\mathbf{P}$ is the 3×1 time-varying heart vector.

Of course, the heart vector and transfer coefficients are unknown for a given individual. However if we had a way to compute this heart vector, we could use it in the solution of the forward problem and obtain the ECG for any body surface location. The approach to solving this problem is based on a physical model of the human torso. The model provides transfer coefficients that relate the potentials at many body surface points to the heart vector. With this information, we select three ECG leads that summarize the intrinsic characteristics of the desired abnormal ECG to simulate. Then we solve the inverse problem to find the cardiac dipole vector

$$\mathbf{P} = \mathbf{B} \times \mathbf{V} \qquad (2.4)$$

where $\mathbf{B}$ is a $3 \times k$ matrix of lead coefficients that is directly derived from inverting the transfer coefficient matrix $\mathbf{T}$. Thus, for the three heart vector components, there are three linear equations of the form

$$p_x(t) = b_{x1}\, v_1(t) + b_{x2}\, v_2(t) + \ldots + b_{xk}\, v_k(t) \qquad (2.5)$$

If we select k body surface ECG leads $\{v_1(t), v_2(t),\ldots, v_k(t)\}$ for which the lead coefficients are known from the physical model of the human torso, we can solve the inverse problem and compute the time-varying heart vector. Once we have these dipole components, we solve the forward problem using Eq. (2.3) to compute the ECG for any point on the body surface.

2.1.3 Genesis of the ECG

Figure 2.7 shows how an ECG is measured using electrodes attached to the body surface and connected to an instrumentation (ECG) amplifier. For the points in time that the vector points toward the electrode connected to the positive terminal of the amplifier, the output ECG signal will be positive-going. If it points to the negative electrode, the ECG will be negative. The time-varying motion of the cardiac vector produces the body surface ECG for one heartbeat with its characteristic P and T waves and QRS complex. Figure 2.8 shows a lead II recording for one heartbeat of a typical normal ECG.

Figure 2.9 illustrates how the cardiac spread of excitation represented by a vector at different points in time relates to the genesis of the body surface ECG for an amplifier configuration like the one in Figure 2.8. In Figure 2.9(a), the slow-moving depolarization of the atria which begins at the sinoatrial (SA) node produces the P wave. As Figure 2.9(b) shows, the signal is delayed in the atrioventricular (AV) node resulting in an isoelectric region after the P wave, then as the Purkinje system starts delivering the stimulus to the ventricular muscle, the onset of the Q wave occurs. In Figure 2.9(c), rapid depolarization of the ventricular muscle is depicted as a large, fast-moving vector which begins producing the R wave. Figure 2.9(d) illustrates that the maximal vector represents a point in time when most of the cells are depolarized, giving rise to the peak of the R wave. In Figure 2.9(e), the final phase of ventricular depolarization occurs as the excitation spreads toward the base of the ventricles (to the top in the picture) giving rise to the S wave.

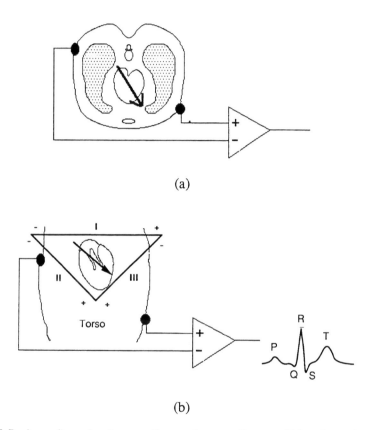

(a)

(b)

Figure 2.7 Basic configuration for recording an electrocardiogram. Using electrodes attached to the body, the ECG is recorded with an instrumentation amplifier. (a) Transverse (top) view of a slice of the body showing the heart and lungs. (b) Frontal view showing electrodes connected in an approximate lead II configuration.

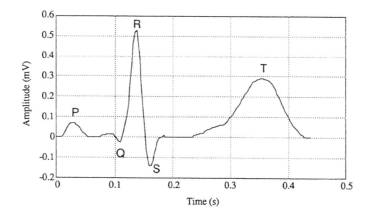

Figure 2.8 Electrocardiogram (ECG) for one normal heartbeat showing typical amplitudes and time durations for the P, QRS, and T waves.

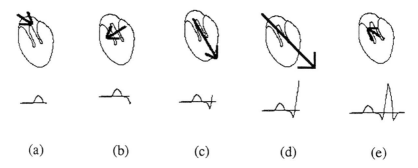

(a)   (b)   (c)   (d)   (e)

Figure 2.9 Relationship between the spread of cardiac electrical activation represented at various time instants by a summing vector (in the upper frames) and the genesis of the ECG (in the lower frames).

2.1.4 The standard limb leads

Figure 2.10 shows how we can view the potential differences between the limbs as ideal voltage sources since we make each voltage measurement using an instrumentation amplifier with a very high input impedance. It is clear that these three voltages form a closed measurement loop. From Kirchhoff's voltage law, the sum of the voltages around a loop equals zero. Thus

$$II - I - III = 0 \qquad\qquad (2.6)$$

We can rewrite this equation to express any one of these leads in terms of the other two leads.

$$II = I + III \qquad\qquad (2.7a)$$

$$I = II - III \qquad\qquad (2.7b)$$

$$III = II - I \qquad\qquad (2.7c)$$

It is thus clear that one of these voltages is completely redundant; we can measure any two and compute the third. In fact, that is exactly what modern ECG machines do. Most machines measure leads I and II and compute lead III. You might ask why we even bother with computing lead III; it is redundant so it has no new information not contained in leads I and II. For the answer to this question, we need to go back to Figure 2.1 and recall that cardiologists learned the relationships between diseases and ECGs by looking at a standard set of leads and relating the appearance of each to different abnormalities. Since these three leads were selected in the beginning, the appearance of each of them is important to the cardiologist.

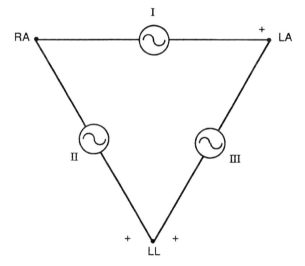

Figure 2.10 Leads I, II, and III are the potential differences between the limbs as indicated. RA and LA are the right and left arms and LL is the left leg.

2.1.5 The augmented limb leads

The early instrumentation had inadequate gain to produce large enough ECG traces for all subjects, so the scheme in Figure 2.11 was devised to produce larger amplitude signals. In this case, the left arm signal, called augmented limb lead aVL, is measured using the average of the potentials on the other two limbs as a reference.

We can analyze this configuration using standard circuit theory. From the bottom left loop

$$i \times R + i \times R - \mathrm{II} = 0 \tag{2.8}$$

or

$$i \times R = \frac{\mathrm{II}}{2} \tag{2.9}$$

From the bottom right loop

$$-i \times R + \mathrm{III} + \mathrm{aVL} = 0 \tag{2.10}$$

or

$$\mathrm{aVL} = i \times R - \mathrm{III} \tag{2.11}$$

Combining Eqs. (2.9) and (2.11) gives

$$\mathrm{aVL} = \frac{\mathrm{II}}{2} - \mathrm{III} = \frac{\mathrm{II} - 2 \times \mathrm{III}}{2} \tag{2.12}$$

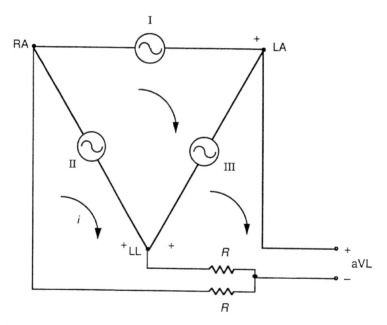

Figure 2.11 The augmented limb lead aVL is measured as shown.

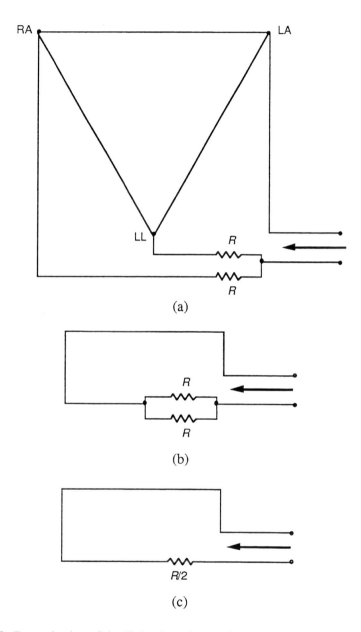

(a)

(b)

(c)

Figure 2.12 Determination of the Thévenin resistance for the aVL equivalent circuit. (a) All ideal voltage sources are shorted out. (b) This gives rise to the parallel combination of two equal resistors. (c) The Thévenin equivalent resistance thus has a value of $R/2$.

From the top center loop

$$II = III + I \qquad (2.13)$$

Substituting gives

$$aVL = \frac{III + I - 2 \times III}{2} = \frac{I - III}{2} \qquad (2.14)$$

This is the Thévenin equivalent voltage for the augmented lead aVL as an average of two of the frontal limb leads. It is clear that aVL is a redundant lead since it can be expressed in terms of two other leads. The other two augmented leads, aVR and aVF, similarly can both be expressed as functions of leads I and III. Thus here we find an additional three leads, all of which can be calculated from two of the frontal leads and thus are all redundant with no new real information. However due to the empirical nature of electrocardiology, the physician nonetheless still needs to see the appearance of these leads to facilitate the diagnosis.

Figure 2.12 shows how the Thévenin equivalent resistance is found by shorting out the ideal voltage sources and looking back from the output terminals.

Figure 2.13 illustrates that a recording system includes an additional resistor of a value equal to the Thévenin equivalent resistance connected to the positive input of the differential instrumentation amplifier. This balances the resistance at each input of the amplifier in order to ensure an optimal common mode rejection ratio (CMRR.

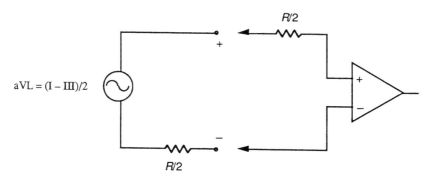

Figure 2.13 In a practical device for recording aVL, a resistance equal to the Thévenin equivalent resistance value of $R/2$ is added at the positive terminal of the instrumentation amplifier to balance the impedance on each input of the amplifier. This is done for optimal common mode performance.

Figure 2.14 shows how to solve vectorially for an augmented limb lead in terms of two of the standard limb leads. The limb leads are represented by vectors oriented in the directions of their corresponding triangle sides but centered at a common origin. To find aVL as in this example, we use the vectors of the two limb leads that connect to the limb being measured, in this case, the left arm. We use lead I as one of the vectors to sum since its positive end connects to the left arm. We negate the vector for limb lead III (i.e., rotate it 180°) since its negative end connects to the left arm. Lead aVL is half the vector sum of leads I and −III [see Eq. (2.14)].

Figure 2.15 shows the complete set of vectors representing the frontal limb leads. From this depiction, you can quickly find all three augmented leads as functions of the frontal leads.

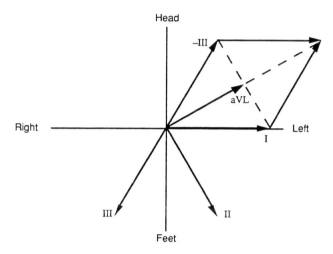

Figure 2.14 The vector graph solution for aVL in terms of leads I and III.

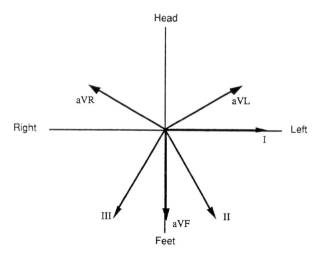

Figure 2.15 The vector relationships among all frontal plane leads.

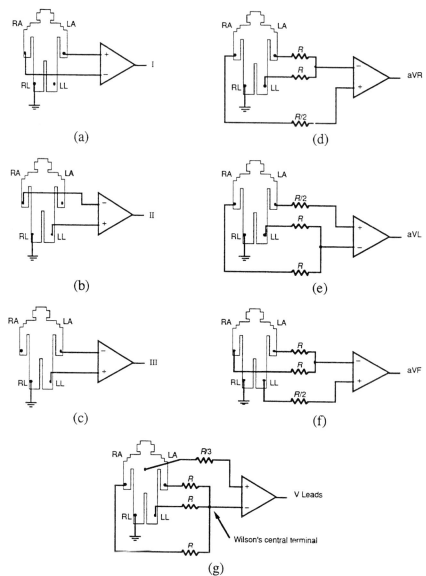

Figure 2.16 Standard 12-lead clinical electrocardiogram. (a) Lead I. (b) Lead II. (c) Lead III. Note the amplifier polarity for each of these limb leads. (d) aVR. (e) aVL. (f) aVF. These augmented leads require resistor networks which average two limb potentials while recording the third. (g) The six V leads are recorded referenced to Wilson's central terminal which is the average of all three limb potentials. Each of the six leads labeled V1–V6 are recorded from a different anatomical site on the chest.

2.2 ECG LEAD SYSTEMS

There are three basic lead systems used in cardiology. The most popular is the 12-lead approach, which defines the set of 12 potential differences that make up the standard clinical ECG. A second lead system designates the locations of electrodes for recording the VCG. Monitoring systems typically analyze one or two leads.

2.2.1 12-lead ECG

Figure 2.16 shows how the 12 leads of the standard clinical ECG are recorded, and Figure 2.17 shows the standard 12-lead ECG for a normal patient. The instrumentation amplifier is a special design for electrocardiography like the one shown in Figure 2.23. In modern microprocessor-based ECG machines, there are eight similar ECG amplifiers which simultaneously record leads I, II, and V1–V6. They then compute leads III, aVL, aVR, and aVF for the final report.

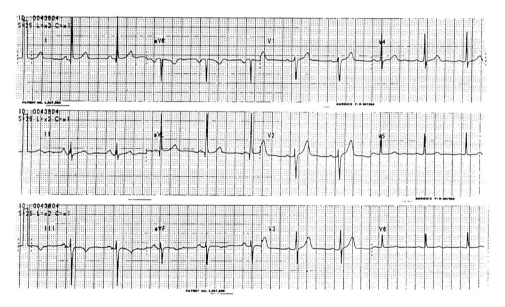

Figure 2.17 The 12-lead ECG of a normal male patient. Calibration pulses on the left side designate 1 mV. The recording speed is 25 mm/s. Each minor division is 1 mm, so the major divisions are 5 mm. Thus in lead I, the R-wave amplitude is about 1.1 mV and the time between beats is almost 1 s (i.e., heart rate is about 60 bpm).

2.2.2 The vectorcardiogram

Figure 2.18 illustrates the placement of electrodes for a Frank VCG lead system. Worldwide this is the most popular VCG lead system. Figure 2.19 shows how potentials are linearly combined with a resistor network to compute the three time-varying orthogonal scalar leads of the Frank lead system. Figure 2.20 is an IBM PC screen image of the VCG of a normal patient.

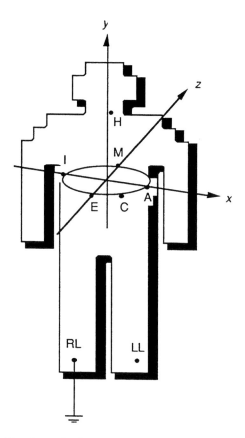

Figure 2.18 The electrode placement for the Frank VCG lead system.

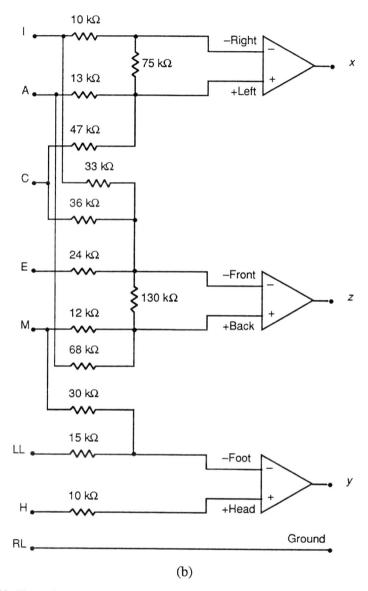

(b)

Figure 2.19 The resistor network for combining body surface potentials to produce the three time-varying scalar leads of the Frank VCG lead system.

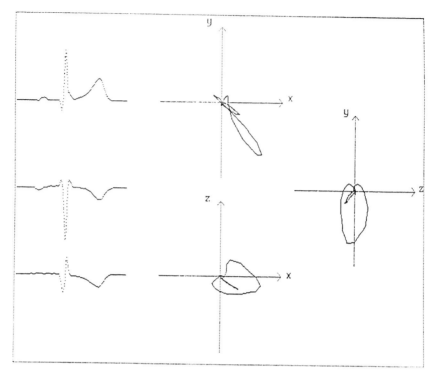

Figure 2.20 The vectorcardiogram of a normal male patient. The three time-varying scalar leads for one heartbeat are shown on the left and are the x, y, and z leads from top to bottom. In the top center is the frontal view of the tip of the vector as it moves throughout one complete heartbeat. In bottom center is a transverse view of the vector loop looking down from above the patient. On the far right is a left sagittal view looking toward the left side of the patient.

2.2.3 Monitoring lead systems

Monitoring applications do not use standard electrode positions but typically use two leads. Since the principal goal of these systems is to reliably recognize each heartbeat and perform rhythm analysis, electrodes are placed so that the primary ECG signal has a large R-wave amplitude. This ensures a high signal-to-noise ratio for beat detection. Since Lead II has a large peak amplitude for many patients, this lead is frequently recommended as the first choice of a primary lead by many manufacturers. A secondary lead with different electrode placements serves as a backup in case the primary lead develops problems such as loss of electrode contact.

2.3 ECG SIGNAL CHARACTERISTICS

Figure 2.21 shows three bandwidths used for different applications in electrocardiography (Tompkins and Webster, 1981). The clinical bandwidth used for recording the standard 12-lead ECG is 0.05–100 Hz. For monitoring applications, such as for intensive care patients and for ambulatory patients, the bandwidth is restricted to 0.5–50 Hz. In these environments, rhythm disturbances (i.e., arrhythmias) are principally of interest rather than subtle morphological changes in the waveforms. Thus the restricted bandwidth attenuates the higher frequency noise caused by muscle contractions (electromyographic or EMG noise) and the lower frequency noise caused by motion of the electrodes (baseline changes). A third bandwidth used for heart rate meters (cardiotachometers) maximizes the signal-to-noise ratio for detecting the QRS complex. Such a filter passes the frequencies of the QRS complex while rejecting noise including non-QRS waves in the signal such as the P and T waves. This filter helps to detect the QRS complexes but distorts the ECG so much that the appearance of the filtered signal is not clinically acceptable. One other application not shown extends the bandwidth up to 500 Hz in order to measure late potentials. These are small higher-frequency events that occur in the ECG following the QRS complex.

The peak amplitude of an ECG signal is in the range of 1 mV, so an ECG amplifier typically has a gain of about 1,000 in order to bring the peak signal into a range of about 1 V.

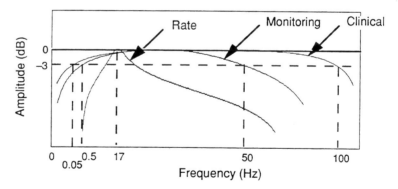

Figure 2.21 Bandwidths used in electrocardiography. The standard clinical bandwidth for the 12-lead clinical ECG is 0.05–100 Hz. Monitoring systems typically use a bandwidth of 0.5–50 Hz. Cardiotachometers for heart rate determination of subjects with predominantly normal beats use a simple bandpass filter centered at 17 Hz and with a Q of about 3 or 4.

2.4 LAB: ANALOG FILTERS, ECG AMPLIFIER, AND QRS DETECTOR*

In this laboratory you will study the characteristics of four types of analog filters: low-pass, high-pass, bandpass and bandstop. You will use these filters to build an ECG amplifier. Next you will study the application of a bandpass filter in a QRS detector circuit, which produces a pulse for each occurrence of a QRS complex. Note that you have to build all the circuits yourself.

2.4.1 Equipment

1. Dual trace oscilloscope
2. Signal generator
3. ECG electrodes
4. Chart recorder
5. Your ECG amplifier and QRS detection board
6. Your analog filter board

2.4.2 Background information

Low-pass filter/integrator

Figure 2.22(a) shows the circuit for a low-pass filter. The low-frequency gain, A_L, is given by

$$A_L = -\frac{R_2}{R_1} \qquad (2.15)$$

The negative sign results because the op amp is in an inverting amplifier configuration. The high-corner frequency is given by

$$f_h = \frac{1}{2\pi R_2 C_1} \qquad (2.16)$$

A low-pass filter acts like an integrator at high frequencies. The integrator output is given by

$$V_0 = -\frac{1}{1 + j\omega R_1 C_1} V_i$$
$$= -\frac{1}{R_1 C_1}\int V_i \, dt \qquad (2.17)$$

---

* Section 2.4 was written by Pradeep Tagare.

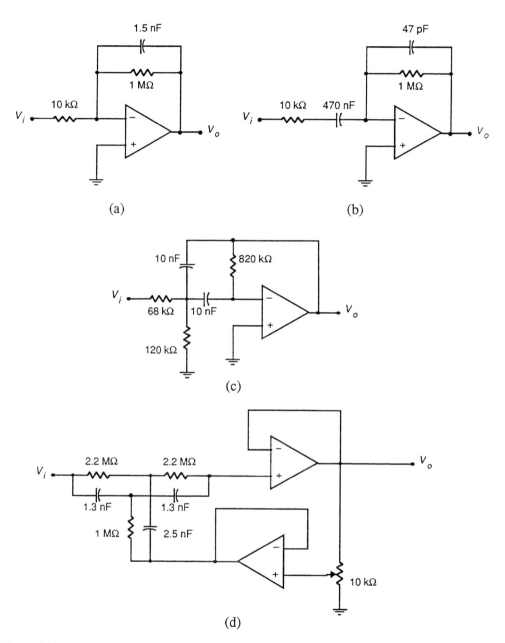

Figure 2.22 Analog filters. (a) Low-pass filter (integrator). (b) High-pass filter (differentiator). (c) Bandpass filter. (d) Bandstop (notch) filter.

This can be verified by observing the phase shift of the output with respect to the input. For a sinusoidal signal, the output is shifted by 90°

$$\int v \sin\omega t = -\frac{v}{\omega} \cos\omega t = \frac{v}{\omega} \sin(\omega t + \frac{\pi}{2}) \tag{2.18}$$

Thus the gain of the integrator falls at high frequencies. Also note that if R_2 were not included in the integrator, the gain would become infinite at dc. Thus at dc the op amp dc bias current charges the integrating capacitor C_1 and saturates the amplifier.

High-pass filter/differentiator

In contrast to the low-pass filter which acts as an integrator at high frequencies, the high-pass filter acts like a differentiator at low frequencies. Referring to Figure 2.22(b), we get the high-frequency gain A_h and the low-corner frequency f_L as

$$A_h = -\frac{R_2}{R_1} \tag{2.19}$$

$$f_L = \frac{1}{2\pi R_1 C_1} \tag{2.20}$$

The differentiating behavior of the high-pass filter at low frequencies can be verified by deriving equations as was done for the integrator. Capacitor C_2 is added to improve the stability of the differentiator. The differentiator gain increases with frequency, up to the low-corner frequency.

Bandpass filter

The circuit we will use is illustrated in Figure 2.22(c). The gain of a bandpass filter is maximum at the center frequency and falls off on either side of the center frequency. The bandwidth of a bandpass filter is defined as the difference between the two corner frequencies. The Q of a bandpass filter is defined as

$$Q = \frac{\text{center frequency}}{\text{bandwidth}} \tag{2.21}$$

Bandstop/notch filter

Line frequency noise is a major source of interference. Sometimes a 60-Hz bandstop (notch) filter is used to reject this interference. Basically such a filter rejects one particular frequency while passing all other frequencies. Figure 2.22(d) shows the bandstop filter that we will use. For the 60-Hz notch filter shown, the 60-Hz rejection factor is defined as

$$\text{60-Hz rejection factor} = \frac{\text{output voltage of the filter at 100 Hz}}{\text{output voltage of the filter at 60 Hz}} \quad (2.22)$$

for the same input voltage.

ECG amplifier

An ECG signal is usually in the range of 1 mV in magnitude and has frequency components from about 0.05–100 Hz. To process this signal, it has to be amplified. Figure 2.23 shows the circuit of an ECG amplifier. The typical characteristics of an ECG amplifier are high gain (about 1,000), 0.05–100 Hz frequency response, high input impedance, and low output impedance. Derivation of equations for the gain and frequency response are left as an exercise for the reader.

QRS detector

Figures 2.24 and 2.25 show the block diagram and complete schematic for the QRS detector. The QRS detector consists of the following five units:

1. QRS filter. The power spectrum of a normal ECG signal has the greatest signal-to-noise ratio at about 17 Hz. Therefore to detect the QRS complex, the ECG is passed through a bandpass filter with a center frequency of 17 Hz and a bandwidth of 6 Hz. This filter has a large amount of ringing in its output.
2. Half-wave rectifier. The filtered QRS is half-wave rectified, to be subsequently compared with a threshold voltage generated by the detector circuit.
3. Threshold circuit. The peak voltage of the rectified and filtered ECG is stored on a capacitor. A fraction of this voltage (threshold voltage) is compared with the filtered and rectified ECG output.
4. Comparator. The QRS pulse is detected when the threshold voltage is exceeded. The capacitor recharges to a new threshold voltage after every pulse. Hence a new threshold determined from the past history of the signal is generated after every pulse.
5. Monostable. A 200-ms pulse is generated for every QRS complex detected. This pulse drives a LED.

Some patients have a cardiac pacemaker. Since sharp pulses of the pacemaker can cause spurious QRS pulse detection, a circuit is often included to reject pacemaker pulses. The rejection is achieved by limiting the slew rate of the amplifier.

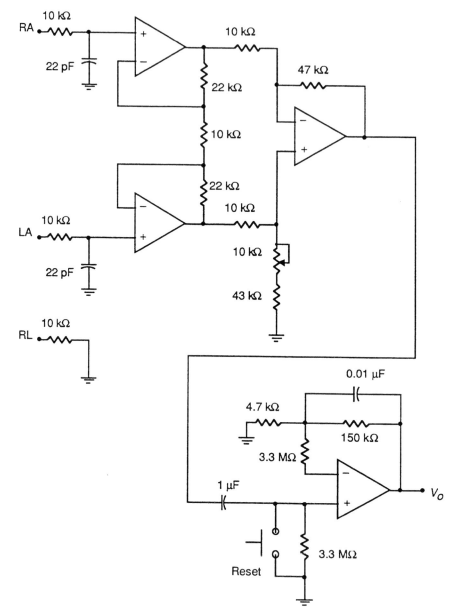

Figure 2.23 Circuit diagram of an ECG amplifier.

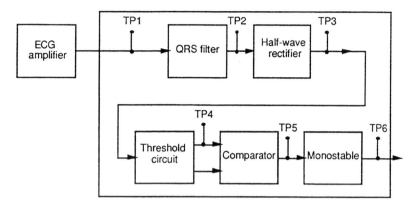

Figure 2.24 Block diagram of a QRS detector.

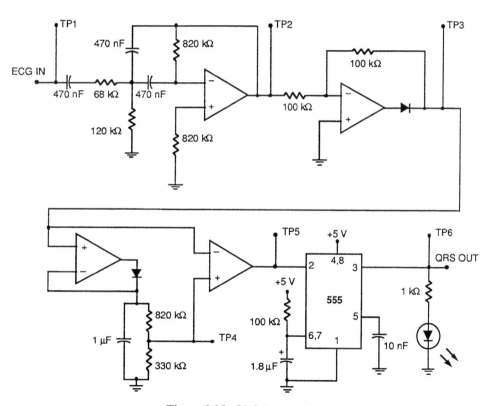

Figure 2.25 QRS detector circuit.

2.4.3 Experimental Procedure

Build all the circuits described above using the LM324 quad operational amplifier integrated circuit shown in Figure 2.26.

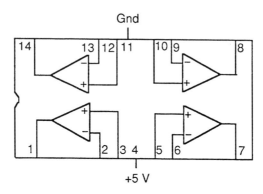

Figure 2.26 Pinout of the LM324 quad operational amplifier integrated circuit.

Low-pass filter

1. Turn on the power to the filter board. Feed a sinusoidal signal of the least possible amplitude generated by the signal generator at 10 Hz into the integrator input and observe both the input and the output on the oscilloscope. Calculate the gain.
2. Starting with a frequency of 10 Hz, increase the signal frequency in steps of 10 Hz up to 200 Hz and record the output at each frequency. You will use these values to plot a graph of the output voltage versus frequency. Next, find the generator frequency for which the output is 0.707-times that observed at 10 Hz. This is the −3 dB point or the high-corner frequency. Record this value.
3. Verify the operation of a low-pass filter as an integrator at high frequencies by observing the phase shift between the input and the output. Record the phase shift at the high-corner frequency.

High-pass filter

1. Feed a sinusoidal signal of the least possible amplitude generated by the signal generator at 200 Hz into the differentiator input and observe both the input and the output on the oscilloscope. Calculate the gain.
2. Starting with a frequency of 200 Hz, decrease the signal frequency in steps of 20 Hz to near dc and record the output at each frequency. You will use these values to plot a graph of the output voltage versus frequency. Next find the gen-

erator frequency for which the output is 0.707-times that observed at 200 Hz. This is the 3 dB point or the low-corner frequency. Record this value.
3. Verify the operation of a high-pass filter as a differentiator at low frequencies by observing the phase shift between the input and the output. Record the phase shift at the low-corner frequency. Another simple way to observe the differentiating behavior is to feed a 10-Hz square wave into the input and observe the spikes at the output.

Bandpass filter

For a 1-V p-p sinusoidal signal, vary the frequency from 10–150 Hz. Record the high- and low-corner frequencies. Find the center frequency and the passband gain of this filter.

Bandstop/notch filter

Feed a 1-V p-p 60-Hz sinusoidal signal into the filter, and measure the output voltage. Repeat the same for a 100-Hz sinusoid. Record results.

ECG amplifier

1. Connect LA and RA inputs of the amplifier to ground and observe the output. Adjust the 100-kΩ pot to null the offset voltage.
2. Connect LA and RA inputs to the signal high and the RL input to signal high (60 Hz) and RL to signal low. This is the common mode operation. Calculate the common mode gain.
3. Connect the LA input to the signal high (30 Hz) and the RA input to the signal low (through an attenuator to avoid saturation). This is the differential mode operation. Calculate the differential mode gain.
4. Find the frequency response of the amplifier.
5. Connect three electrodes to your body. Connect these electrodes to the amplifier inputs. Observe the amplifier output. If the signal is very noisy, try twisting the leads together. When you get a good signal, get a recording on the chart recorder.

QRS detector

1. Apply three ECG electrodes. Connect the electrodes to the input of the ECG amplifier board. Turn on the power to the board and observe the output of the ECG amplifier on the oscilloscope. Try pressing the electrodes if there is excessive noise.
2. Connect the output of the ECG amplifier to the input of the QRS detector board. Observe the following signals on the oscilloscope and then record them on a stripchart recorder with the ECG (TP1) on one channel and each of the other test signals (TP2–TP6) on the other channel. Use a reasonably fast paper speed (e.g., 25 mm/s).

Signals to be observed:

Test point	Signal
TP1	Your ECG
TP2	Filtered output
TP3	Rectified output
TP4	Comparator input
TP5	Comparator output
TP6	Monostable output

The LED should flash for every QRS pulse detected.

2.4.4 Lab report

1. Using equations described in the text, determine the values of A_L and f_h for the low-pass filter. Compare these values with the respective values obtained in the lab and account for any differences.
2. Plot the graph of the filter output voltage versus frequency. Show the −3 dB point on this graph.
3. What value of phase shift did you obtain for the low-pass filter?
4. Using equations described in the text, determine the values of A_H and f_L for the high-pass filter. Compare these values with the respective values obtained in the lab and account for any differences.
5. Plot the graph of the filter output voltage versus frequency. Show the −3 dB point on this graph.
6. What value of phase shift did you obtain for the high-pass filter?
7. For the bandpass amplifier, list the values that you got for the following:
   (a) center frequency
   (b) passband gain
   (c) bandwidth
   (d) Q
   Show all calculations.
8. What is the 60-Hz rejection factor for the bandstop filter you used?
9. What are the upper and lower −3 dB frequencies of your ECG amplifier? How do they compare with the theoretical values?
10. What is the gain of your ECG amplifier? How does it compare with the theoretical value?
11. What is the CMRR of your ECG amplifier?
12. How would you change the −3 dB frequencies of this amplifier?
13. Explain the waveforms you recorded on the chart recorder. Are these what you would expect to obtain?
14. Will the QRS detector used in this lab work for any person's ECG? Justify your answer.

Include all chart recordings with your lab report and show calculations wherever appropriate.

2.5 REFERENCES

Tompkins, W. J. and Webster, J. G. (eds.) 1981. *Design of Microcomputer-based Medical Instrumentation*. Englewood Cliffs, NJ: Prentice-Hall.

2.6 STUDY QUESTIONS

2.1   What is a cardiac equivalent generator? How is it different from the actual cardiac electrical activity? Give two examples.

2.2   What is the vectorcardiogram and how is it recorded?

2.3   The heart vector of a patient is oriented as shown below at one instant of time. At this time, which of the frontal leads (I, II, and III) are positive-going for:

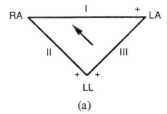 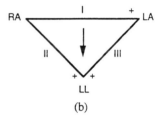

(a)   (b)

2.4   A certain microprocessor-based ECG machine samples and stores only leads I and II. What other standard leads can it compute from these two?

2.5   It is well known that all six frontal leads of the ECG can be expressed in terms of any two of them. Express the augmented lead at the right arm (i.e., aVR) in terms of leads I and II.

2.6   Express Lead II in terms of aVF and aVL.

2.7   Is it possible to express lead V6 in terms of two other leads? Is there any way to calculate V6 from a larger set of leads?

2.8   There are four different bandwidths that are used in electrocardiography. Describe the principal applications for each of these bandwidths.

2.9   What is the frequency range of the standard 3-dB bandwidth used in (a) clinical electrocardiography, (b) electrocardiography monitoring applications such as in the intensive care unit? (c) Why are the clinical and monitoring bandwidths different?

2.10 A cardiologist records a patient's ECG on a machine that is suspected of being defective. She notices that the QRS complex of a normal patient's ECG has a lower peak-to-peak amplitude than the one recorded on a good machine. Explain what problems in instrument bandwidth might be causing this result.

2.11 A cardiologist notices that the T wave of a normal patient's ECG is distorted so that it looks like a biphasic sine wave instead of a unipolar wave. Explain what problems in instrument bandwidth might be causing this problem.

2.12 What is the electrode material that is best for recording the ECG from an ambulatory patient?

2.13 A cardiotachometer uses a bandpass filter to detect the QRS complex of the ECG. What is its center frequency (in Hz)? How was this center frequency determined?

2.14 An engineer designs a cardiotachometer that senses the occurrence of a QRS complex with a simple amplitude threshold. It malfunctions in two patients. (a) One patient's ECG has baseline drift and electromyographic noise. What ECG preprocessing step would provide the most effective improvement in the design for this case? (b) Another patient has a T wave that is much larger than the QRS complex. This false triggers the thresholding circuit. What ECG preprocessing step would provide the most effective improvement in the design for this case?

2.15 What is included in the design of an averaging cardiotachometer that prevents it from responding instantaneously to a heart rate change?

2.16 A typical modern microprocessor-based ECG machine samples and stores leads I, II, V1, V2, V3, V4, V5, and V6. From this set of leads, calculate (a) lead III, (b) augmented lead aVF.

3

Signal Conversion

David J. Beebe

The power of the computer to analyze and visually represent biomedical signals is of little use if the analog biomedical signal cannot be accurately captured and converted to a digital representation. This chapter discusses basic sampling theory and the fundamental hardware required in a typical signal conversion system. Section 3.1 discusses sampling basics in a theoretical way, and section 3.2 describes the circuits required to implement a real signal conversion system. We examine the overall system requirements for an ECG signal conversion system and discuss the possible errors involved in the conversion. We review digital-to-analog and analog-to-digital converters and other related circuits including amplifiers, sample-and-hold circuits, and analog multiplexers.

3.1 SAMPLING BASICS*

The whole concept of converting a continuous time signal to a discrete representation usable by a microprocessor, lies in the fact that we can represent a continuous time signal by its instantaneous amplitude values taken at periodic points in time. More important, we are able to reconstruct the original signal perfectly with just these sampled points. Such a concept is exploited in movies, where individual frames are snapshots of a continuously changing scene. When these individual frames are played back at a sufficiently fast rate, we are able to get an accurate representation of the original scene (Oppenheim and Willsky, 1983).

3.1.1 Sampling theorem

The *sampling theorem* initially developed by Shannon, when obeyed, guarantees that the original signal can be reconstructed from its samples without any loss of information. It states that, for a continuous <u>bandlimited</u> signal that contains no fre-

---

* Section 3.1 was written by Annie Foong.

quency components higher than f_c, the original signal can be completely recovered without distortion if it is sampled at a rate of at least $2 \times f_c$ samples/s. A sampling frequency f_s of twice the highest frequency *present* in a signal is called the Nyquist frequency.

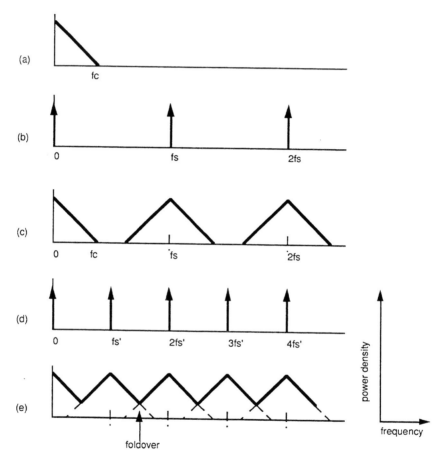

Figure 3.1 Effect in the frequency domain of sampling in the time domain. (a) Spectrum of original signal. (b) Spectrum of sampling function. (c) Spectrum of sampled signal with $f_s > 2f_c$. (d) Spectrum of sampling function with $f_s' < 2f_c$. (e) Spectrum of sampled signal with $f_s' < 2f_c$.

3.1.2 Aliasing, foldover, and other practical considerations

To gain more insight into the mechanics of sampling, we shall work in the frequency domain and deal with the spectra of signals. As illustrated in Figure 3.1(c), if we set the sampling rate larger than $2 \times f_c$, the original signal can be recovered by placing a low-pass filter at the output of the sampler. If the sampling rate is too low, the situation in Figure 3.1(e) arises. Signals in the overlap areas are *dirtied* and cannot be recovered. This is known as *aliasing* where the higher frequencies are reflected into the lower frequency range (Oppenheim and Willsky, 1983).

Therefore, if we know in advance the highest frequency in the signal, we can theoretically establish the sampling rate at twice the highest frequency present. However, real-world signals are corrupted by noise that frequently contains higher frequency components than the signal itself. For example, if an ECG electrode is placed over a muscle, an unwanted electromyographic (EMG) signal may also be picked up (Cromwell et al., 1976). This problem is usually minimized by placing a low-pass filter at the sampler's input to keep out the unwanted frequencies.

However, nonideal filters can also prevent us from perfectly recovering the original signal. If the input filter is not ideal as in Figure 3.2(a), which is the case in practice, high frequencies may still slip through, and aliasing may still be present.

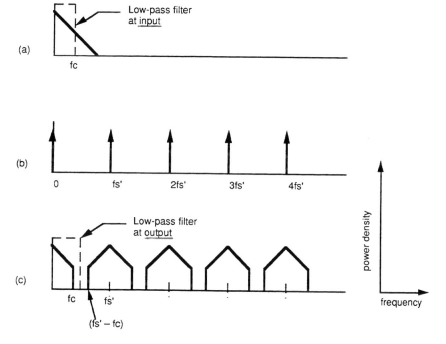

Figure 3.2 Effects of input and output filters. (a) Low-pass filter at input to remove high frequencies of signal. (b) Spectrum of sampling function. (c) No foldover present, low-pass filter at output to recover original signal.

Often ignored is the effect of the output filter. It can be seen in Figure 3.2(c) that, if the output filter is nonideal, the reconstructed signal may not be correct. In particular, the cutoff frequency of the output filter must be larger than f_c, but smaller than $(f_s' - f_c)$ so as not to include undesired components from the next sequence of spectra.

Finally, we may be limited by practical considerations not to set the sampling rate at Nyquist frequency even if we do know the highest frequency present in a signal. Most biomedical signals are in the low-frequency range. Higher sampling rates require more expensive hardware and larger storage space. Therefore, we usually tolerate an acceptable level of error in exchange for more practical sampling rates.

3.1.3 Examples of biomedical signals

Figure 3.3 gives the amplitudes and frequency bands of several human physiological signals.

Electroencephalogram (EEG)
     Frequency range: dc–100 Hz (0.5–60 Hz)
     Signal range: 15–100 mV

Electromyogram (EMG)
     Frequency range: 10–200 Hz
     Signal range: function of muscle activity and electrode placement

Electrocardiogram (ECG)
     Frequency range: 0.05–100 Hz
     Signal range: 10 μV(fetal), 5 mV(adult)

Heart rate
     Frequency range: 45–200 beats/min

Blood pressure
     Frequency range: dc–200 Hz (dc–60 Hz)
     Signal range: 40–300 mm Hg (arterial); 0–15 mm Hg (venous)

Breathing rate
     Frequency range: 12–40 breaths/min

Figure 3.3 Biomedical signals and ranges; major diagnostic range shown in brackets.

3.2 SIMPLE SIGNAL CONVERSION SYSTEMS

Biomedical signals come in all shapes and sizes. However, to capture and analyze these signals, the same general processing steps are required for all the signals. Figure 3.4 illustrates a general analog-to-digital (A/D) signal conversion system.

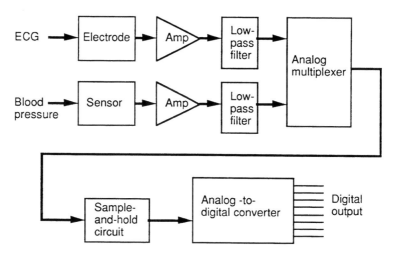

Figure 3.4 A typical analog-to-digital signal conversion system consists of sensors, amplifiers, multiplexers, a low-pass filter, a sample-and-hold circuit, and the A/D converter.

First the signal must be captured. If it is electrical in nature, a simple electrode can be used to pass the signal from the body to the signal conversion system. For other signals, a sensor is required to convert the biomedical signal into a voltage. The signal from the electrode or sensor is usually quite small in amplitude (e.g., the ECG ranges from 10 µV to 5 mV). Amplification is necessary to bring the amplitude of the signal into the range of the A/D converter. The amplification should be done as close to the signal source as possible to prevent any degradation of the signal. If there are several input signals to be converted, an analog multiplexer is needed to route each signal to the A/D converter. In order to minimize aliasing, a low-pass filter is often used to bandlimit the signal prior to sampling. A sample-and-hold circuit is required (except for very slowly changing signals) at the input to the A/D converter to hold the analog signal at a constant value during the conversion process. Finally, the A/D converter changes the analog voltage stored by the sample-and-hold circuit to a digital representation.

Now that a digital version of the biomedical signal has been obtained, what can it be used for? Often the digital information is stored in a memory device for later processing by a computer. The remaining chapters discuss in detail a variety of

digital signal processing algorithms commonly used for processing biomedical signals. In some cases this processing can be done in real time. Another possible use
for the digital signal is in a control system. In this case the signal is processed by a
computer and then fed back to the device to be controlled. Often the controller requires an analog signal, so a digital-to-analog (D/A) converter is needed. Figure 3.5
illustrates a general D/A conversion system. The analog output might be used to
control the flow of gases in an anesthesia machine or the temperature in an
incubator.

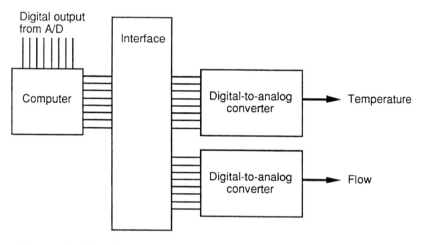

Figure 3.5 A typical digital-to-analog signal conversion system with a computer for processing
the digital signal prior to conversion.

3.3 CONVERSION REQUIREMENTS FOR BIOMEDICAL SIGNALS

As discussed in section 3.1.3, biomedical signals have a variety of characteristics.
The ultimate goal of any conversion system is to convert the biomedical signal to a
digital representation with a minimal loss of information. The specifications for
any conversion system are dependent on the signal characteristics and the application. In general, from section 3.1.3, one can see that biomedical signals are typically low frequency and low amplitude in nature. The following attributes should
be considered when designing a conversion system: (1) accuracy, (2) sampling
rate, (3) gain, (4) processing speed, (5) power consumption, and (6) size.

3.4 SIGNAL CONVERSION CIRCUITS

The digital representation of a continuous analog signal is discrete in both time (determined by the sampling rate) and amplitude (determined by the number of bits in a sampled data word). A variety of circuit configurations are available for converting signals between the analog and digital domains. Many of these are discussed in this chapter. Each method has its own advantages and shortcomings. The discussion here is limited to those techniques most commonly used in the conversion of biomedical signals. The D/A converter is discussed first since it often forms part of an A/D converter.

3.4.1 Converter characteristics

Before describing the details of the converter hardware, it is important to gain some knowledge of the basic terminology used in characterizing a converter's performance. One common method of examining the characteristics of a D/A converter (or an A/D converter) is by looking at its static and dynamic properties as described by Allen and Holberg (1987).

Static

For illustrative purposes a D/A converter is used, but the static errors discussed also apply to A/D converters. The ideal static behavior of a 3-bit D/A converter is shown in Figure 3.6. All the combinations of the digital input word are on the horizontal axis, while the analog output is on the vertical axis. The maximum analog output signal is 7/8 of the full scale reference voltage V_{ref}. For each unique digital input there should be a unique analog output. Any deviations from Figure 3.6 are known as static conversion errors. Static conversion errors can be divided into integral linearity, differential linearity, monotonicity, and resolution.

Integral linearity is the maximum deviation of the output of the converter from a straight line drawn from its ideal minimum to its ideal maximum. Integral linearity is often expressed in terms of a percentage of the full scale range or in terms of the least significant bit (LSB). Integral linearity can be further divided into absolute linearity, offset or zero error, full scale error, gain error, and monotonicity errors. Absolute linearity emphasizes the zero and full scale errors. The zero or offset error is the difference between the actual output and zero when the digital word for a zero output is applied. The full scale error is the difference between the actual and the ideal voltage when the digital word for a full scale output is applied. A gain error exists when the slope of the actual output is different from the slope of the ideal output. Figure 3.7(a) illustrates offset and gain errors. Monotonicity in a D/A converter means that as the digital input to the converter increases over its full scale range, the analog output never exhibits a decrease between one conversion step and the next. In other words, the slope of the output is never negative. Figure 3.7(b) shows the output of a converter that is nonmonotonic.

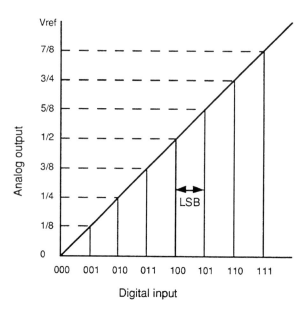

Figure 3.6 The ideal static behavior for a three-bit D/A converter. For each digital word there should be a unique analog signal.

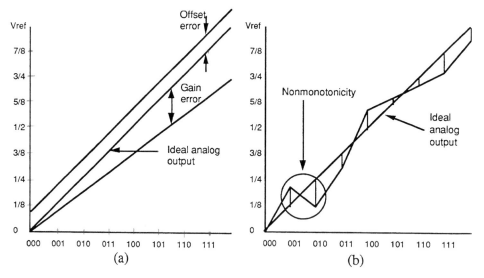

Figure 3.7 Digital-to-analog converter characteristics. (a) Gain and offset errors. (b) Monotonicity errors.

Differential linearity differs from integral linearity in that it is a measure of the separation between adjacent levels (Allen and Holberg, 1987). In other words, differential linearity measures bit-to-bit deviations from the ideal output step size of 1 LSB. Figure 3.8 illustrates the differences between integral and differential linearity.

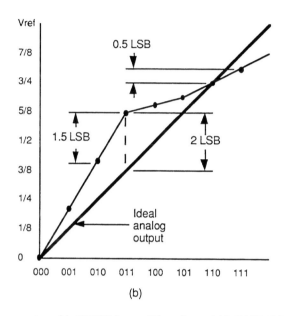

(b)

Figure 3.8 A D/A converter with ±2 LSB integral linearity and ±0.5 LSB differential linearity.

Resolution is defined as the smallest input digital code for which a unique analog output level is produced. Theoretically the resolution of an n-bit D/A converter is 2^n discrete analog output levels. In reality, the resolution is often less due to noise and component drift.

Dynamic

Settling time, for a D/A converter, is defined as the time it takes for the converter to respond to an input change. More explicitly, settling time is the time between the time a new digital signal is received at the converter input and the time when the output has reached its final value (within some specified tolerance). Many factors affect the settling time, so the conditions under which the settling time was found must be clearly stated if comparisons are to be made. Factors to note include the magnitude of the input change applied and the load on the output. Settling time in

D/A converters is important as it relates directly to the speed of the conversion. The output must settle sufficiently before it can be used in any further processing.

3.4.2 Digital-to-analog converters

The objective of the D/A converter is to construct an analog output signal for a given digital input. The first requirement for a D/A converter is an accurate voltage reference. Next the reference must be scaled to provide analog outputs at levels corresponding to each possible digital input. This is usually implemented using either voltage or charge scaling. Finally, the output can be interpolated to provide a smooth analog output.

Voltage reference

An accurate voltage reference is essential to the operation of a D/A converter. The analog output is derived from this reference voltage. Common reference voltage errors are either due to initial adjustments or generated by drifts with time and temperature. Two types of voltage references are used. One type uses the reverse breakdown voltage of a zener diode, while the other type derives its reference voltage from the extrapolated band-gap voltage of silicon. Temperature compensation is used in both cases. Scaling of this reference voltage is usually accomplished with passive components (resistors for voltage scaling and capacitors for charge scaling).

Voltage scaling

Voltage scaling of the reference voltage uses series resistors connected between the reference voltage and ground to selectively obtain discrete voltages between these limits. Figure 3.9 shows a simple voltage scaling 3-bit D/A converter. The digital input is decoded to select the corresponding output voltage. The voltage scaling structure is very regular and uses a small range of resistances. This is well suited to integrated circuit technology. Integrated circuit fabrication is best at making the same structure over and over. So while control over the absolute value of a resistor might be as high as 50 pecent, the relative accuracy can be as low as one percent (Allen and Holberg, 1987). That is, if we fabricate a set of resistors on one piece of silicon, each with a nominal value of 10 kΩ, the value of each might actually be 15 kΩ, but they will all be within one percent of each other in value.

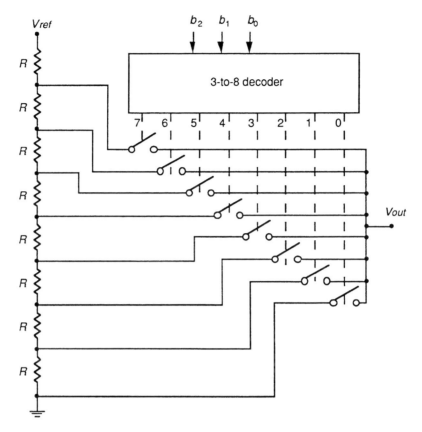

Figure 3.9 A simple voltage scaling D/A converter.

(2) *Charge scaling*

Charge scaling D/A converters operate by doing binary division of the total charge applied to a capacitor array. Figure 3.10 shows a 3-bit charge scaling D/A converter. A two-phase clock is used. During phase 1, S_0 is closed and switches b_2, b_1, b_0 are closed shorting all the capacitors to ground. During phase 2 the capacitors associated with bits that are "1" are connected to V_{ref} and those with bits that are "0" are connected to ground. The output is valid only during phase 2. Equations (3.1) and (3.2) describe this operation. Note that the total capacitance is always $2C$ regardless of the number of bits in the word to be converted. The accuracy of charge scaling converters depends on the capacitor ratios. The error ratio for integrated circuit capacitors is frequently as low as 0.1 percent (Allen and Holberg, 1987).

$$C_{eq} = b_2\, C + b_1\, \frac{C}{2} + b_0\, \frac{C}{4} \qquad\qquad (3.1)$$

$$V_{out} = \frac{C_{eq}}{2C} \qquad\qquad (3.2)$$

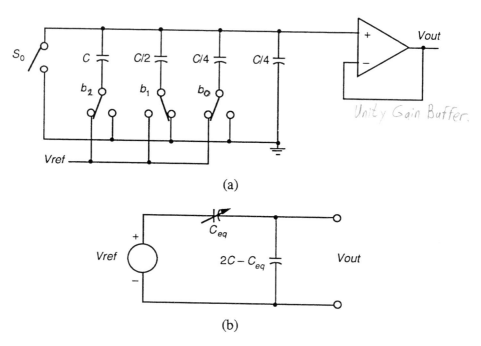

(a)

(b)

Figure 3.10 A 3-bit charge scaling D/A converter. a) Circuit with a binary 101 digital input. b) Equivalent circuit with any digital input.

Output interpolation

The outputs of simple D/A converters, such as those shown in Figures 3.9 and 3.10, are limited to discrete values. Interpolation techniques are often used to reconstruct an analog signal. Interpolation methods that can be easily implemented with electronic circuits include the following techniques: (1) zero-order hold or one-point, (2) linear or two-point, (3) bandlimited or low-pass (Tompkins and Webster, 1981).

3.4.3 Analog-to-digital converters

The objective of an A/D converter is to determine the output digital word for a given analog input. As mentioned previously, A/D converters often make use of D/A converters. Another method commonly used involves some form of integration or ramping. Finally, for high-speed sampling, a parallel or flash converter is used. In most converters a sample-and-hold circuit is needed at the input since it is not possible to convert a changing input signal. For very slowly changing signals, a sample-and-hold circuit is not always required. The errors associated with A/D converters are similar to those described in section 3.4.1 if the input and output definitions are interchanged.

Counter

The counter A/D converter increments a counter to build up the internal output one LSB at a time until it equals the analog input signal. A comparator stops the counter when the internal output has built up to the input signal level. At this point the count equals the digital output. The disadvantage of this scheme is a conversion time that varies with the level of the input signal. So for low-amplitude signals the conversion time can be fast, but if the signal amplitude doubles the conversion time will also double. Also, the accuracy of the conversion is subject to the error in the ramp generation.   Fig 3·11a

Tracking

A variation of the counter A/D converter is the tracking A/D converter. While the counter converter resets its internal output to zero after each conversion, the internal output in the tracking converter continues to follow the analog input. Figure 3.11 illustrates this difference. By externally stopping the tracking A/D converter, it can be used as a sample-and-hold circuit with a digital output. Also by disabling the up or the down control, the tracking converter can be used to find the maximum or minimum value reached by the input signal over a given time period (Tompkins and Webster, 1988).

Dual slope

In a dual-slope converter, the analog input is integrated for a fixed interval of time (T_1). The length of this time is equal to the maximum count of the internal counter. The charge accumulated on the integrator's capacitor during this integration time is proportional to input voltage according to

$$Q = CV \qquad (3.3)$$

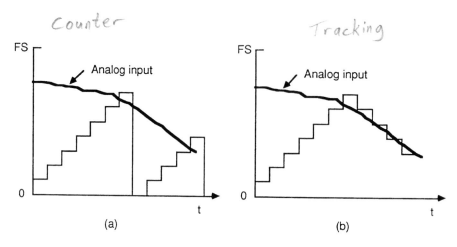

Figure 3.11 Internal outputs of the counter and tracking converters. a) The internal output of a counter A/D converter resets after each conversion. b) The internal output of the tracking A/D converter follows the analog input.

The slope of the integrator output is proportional to the amplitude of the analog input. After time T_1 the input to the integrator is switched to a negative reference voltage V_{ref}, thus the integrator integrates negatively with a constant slope. A counter counts the time t_2 that it takes for the integrator to reach zero. The charge gained by the integrator capacitor during T_1 must equal the charge lost during t_2

$$T_1 V_{in}(avg) = t_2 V_{ref} \qquad (3.4)$$

Note that the ratio of t_2 to T_1 is also the ratio of the counter values

$$\frac{t_2}{T_1} = \frac{V_{in}}{V_{ref}} = \frac{\text{counter}}{\text{fixed count}} \qquad (3.5)$$

So the count at the end of t_2 is equal to the digital output of the A/D converter. Figure 3.12 shows a block diagram of a dual-slope A/D converter and its associated waveforms. Note that the output of the dual-slope A/D converter is not a function of the slope of the integrator nor of the clock rate. As a result, this method of conversion is very accurate.

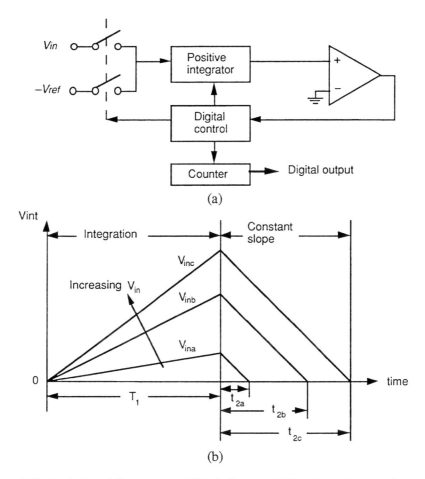

Figure 3.12 Dual-slope A/D converter. a) Block diagram. b) Waveforms illustrate the operation of the converter.

Successive approximation converter

The successive approximation converter uses a combination of voltage-scaling and charge-scaling D/A converters. Figure 3.13 shows a block diagram of a typical successive approximation A/D converter. It consists of a comparator, a D/A converter, and digital control logic. The conversion begins by sampling the analog signal to be converted. Next, the control logic assumes that the MSB is "1" and all other bits are "0". This digital word is applied to the D/A converter and an internal analog signal of 0.5 V_{ref} is generated. The comparator is now used to compare this generated analog signal to the analog input signal. If the comparator output is high, then the MSB is indeed "1". If the comparator output is "0", then the MSB is

changed to "0". At this point the MSB has been determined. This process is repeated for each remaining bit in order.

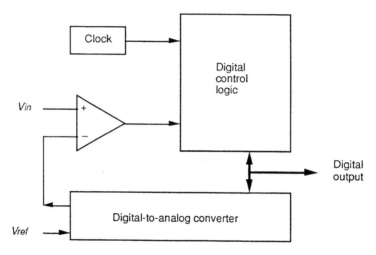

Figure 3.13 Block diagram of a typical successive approximation A/D converter.

Figure 3.14 shows the possible conversion paths for a 3-bit converter. Note that
the number of clock cycles required to convert an n-bit word is n.

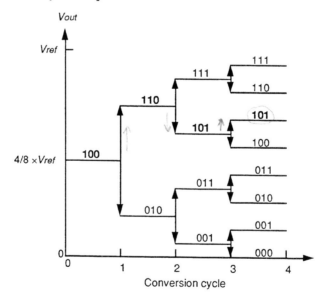

Figure 3.14 The successive approximation process. The path for an analog input equal to
$5/8 \times Vref$ is shown in bold.

Parallel or flash

For very high speed conversions a parallel or flash type converter is used. The ultimate conversion speed is one clock cycle, which would consist of a setup and convert phase. In this type of converter the sample time is often the limiting factor for speed. The operation is straightforward and it is illustrated in Figure 3.15. To convert an n-bit word, $2^n - 1$ comparators are required.

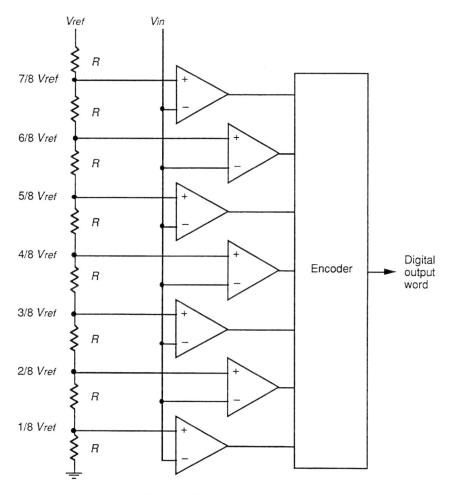

Figure 3.15 A 3-bit flash A/D converter.

3.4.4 Sample-and-hold circuit

Since the conversion from an analog signal to a digital signal takes some finite amount of time, it is advantageous to hold the analog signal at a constant value during this conversion time. Figure 3.16 shows a simple sample-and-hold circuit that can be used to sample the analog signal and freeze its value while the A/D conversion takes place.

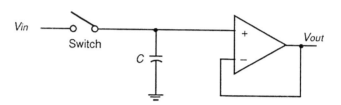

Figure 3.16 A simple implementation of a sample-and-hold circuit.

Errors introduced by the sample-and-hold circuit include offset in the initial voltage storage, amplifier drift, and the slow discharge of the stored voltage. The dynamic properties of the sample-and-hold circuit are important in the overall performance of an A/D converter. The time required to complete one sample determines the minimum conversion time for the A/D. Figure 3.17 illustrates the acquisition time (t_a) and the settling time (t_s).

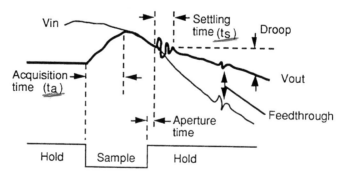

Figure 3.17 Sample-and-hold input and output voltages illustrate specifications.

3.4.5 Analog multiplexer

When several signals need to be converted, it is necessary to either provide an A/D converter for each signal or use an analog multiplexer to direct the various signals to a single converter. For most biomedical signals, the required conversion rates are low enough that multiplexing the signals is the appropriate choice.

Common analog multiplexers utilize either JFET or CMOS transistors. Figure 3.18 shows a simple CMOS analog switch circuit. A number of these switches are connected to a single Vout to make a multiplexer. The switches should operate in a break-before-make fashion to ensure that two input lines are not shorted together. Other attributes to be considered include on-resistance, leakage currents, crosstalk, and settling time.

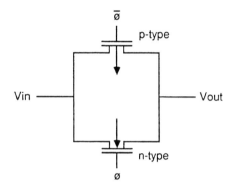

Figure 3.18 A simple CMOS analog switch. The basic functional block of a CMOS analog multiplexer.

3.4.6 Amplifiers

The biomedical signal produced from the sensor or electrode is typically quite small. As a result, the first step before the A/D conversion process is often amplification. Analog amplification circuits can also provide filtering. Analog filters are often used to bandlimit the signal prior to sampling to reduce the sampling rate required to satisfy the sampling theorem and to eliminate noise.

General

For common biomedical signals such as the ECG and the EEG, a simple instrumentation amplifier is used. It provides high input impedance and high CMRR (Common Mode Rejection Ratio). Section 2.4 discusses the instrumentation amplifier and analog filtering in more detail. *P44*

Micropower amplifiers

The need for low-power devices in battery-operated, portable and implantable biomedical devices has given rise to a class of CMOS amplifiers known as micropower devices. Micropower amplifiers operate in the weak-inversion region of transistor operation. This operation greatly reduces the power-supply currents required and also allows for operation on very low supply voltages (1.5 V or even lower). Obviously at such low supply voltages the signal swings must be kept small.

3.5 LAB: SIGNAL CONVERSION

This lab demonstrates the effect of the sampling rate on the frequency spectrum of the signal and illustrates the effects of aliasing in the frequency domain.

3.5.1 Using the Sample utility

1. Load the UW DigiScope program using the directions in Appendix D, select ad(V) Ops, then (S)ample. The Sample menu allows you to read and display a waveform data file, sample the waveform at various rates, display the sampled waveform, recreate a reconstructed version of the original waveform by interpolation, and find the power spectrum of this waveform. The following steps illustrate these functions. You should use this as a tutorial. After completing the tutorial, use the sample functions and go through the lab procedure. Three data files are available for study: a sine wave, sum of sine waves of different frequencies, and a square wave. The program defaults to the single sine wave.
2. The same waveform is displayed on both the upper and lower channels. Since a continuous waveform cannot be displayed, a high sampling rate of 5000 Hz is used. Select (P)wr Spect from the menu to find the spectrum of this waveform. Use the (M)easure option to determine the frequency of the sine wave.
3. Select (S)ample. Type the desired sampling rate in samples/s at the prompt (try 1000) and hit RETURN. The sampled waveform is displayed in the time domain on the bottom channel.
4. Reconstruct the sampled signal using (R)ecreate followed by the zero-order hold (Z)oh option. This shows the waveform as it would appear if you directly displayed the sampled data using a D/A converter.
5. Select (P)wr Spect from the menu to find the spectrum of this waveform. Note that the display runs from 0 to one-half the sampling frequency selected. Use the (M)easure option to determine the predominant frequencies in this waveform. What is the difference in the spectrum of the signal before and after sampling and reconstruction?

6. These steps may be repeated for three other waveforms with the (D)ata Select option.

3.5.2 Procedure

1. Load the sine wave and measure its period. Sample this wave at a frequency much greater than the Nyquist frequency (e.g., 550 samples per second) and reconstruct the waveform using the zero-order hold command. What do you expect the power spectrum of the sampled wave to look like? Perform (P)wr Spect on the sampled data and explain any differences from your expectations. Measure the frequencies at which the peak magnitudes occur.

2. Sample at sampling rates 5–10 percent above, 5–10 percent below, and just at the Nyquist frequency. Describe the appearance of the sampled data and its power spectrum. Measure the frequencies at which peaks in the response occur. Sample the sine wave at the Nyquist frequency several times. What do you notice? Can the signal be perfectly reproduced?

3. Recreate the original data by interpolation using zero-order hold, linear and sinusoidal interpolation. What are the differences between these methods? What do you expect the power spectrum of the recreated data will look like?

4. Repeat the above steps 1–3 using the data of the sum of two sinusoids.

5. Repeat the above steps 1–3 using the square wave data.

3.6 REFERENCES

Allen, P. E. and Holberg, D. R. 1987. *CMOS Analog Circuit Design*. New York: Holt, Rinehart and Winston.

Cromwell, L., Arditti, M., Weibel, F. J., Pfeiffer, E. A., Steele, B., and Labok, J. 1976. *Medical Instrumentation for Health Care*. Englewood Cliffs, NJ: Prentice Hall.

Oppenheim, A. V. and Willsky, A. S. 1983. *Signals and Systems*. Englewood Cliffs, NJ: Prentice Hall.

Tompkins, W. J., and Webster, J. G. (eds.) 1981. *Design of microcomputer-based medical instrumentation*. Englewood Cliffs, NJ: Prentice Hall.

Tompkins, W. J., and Webster, J. G. (eds.) 1988. *Interfacing sensors to the IBM PC*. Englewood Cliffs, NJ: Prentice Hall.

3.7 STUDY QUESTIONS

3.1   What is the purpose of using a low-pass filter prior to sampling an analog signal?

3.2   Draw D/A converter characteristics that illustrate the following errors: (1) offset error of one LSB, (2) integral linearity of ± 1.5 LSB, (3) differential linearity of ± 1 LSB.

3.3   Design a 4-bit charge scaling D/A converter. For $V_{ref} = 5$ V, what is V_{out} for a digital input of 1010?

3.4   Why is a dual-slope A/D converter considered an accurate method of conversion? What will happen to the output if the integrator drifts over time?

3.5 Show the path a 4-bit successive approximation A/D converter will make to converge given an analog input of $9/16\ V_{ref}$.

3.6 Discuss reasons why each attribute listed in section 3.3 is important to consider when designing a biomedical conversion system. For example, size would be important if the device was to be portable.

3.7 List the specifications for an A/D conversion system that is to be used for an EEG device.

3.8 Draw a block diagram of a counter-type A/D converter.

3.9 Explain Shannon's sampling theorem. If only two samples per cycle of the highest frequency in a signal is obtained, what sort of interpolation strategy is needed to reconstruct the signal?

3.10 A 100-Hz-bandwidth ECG signal is sampled at a rate of 500 samples per second. (a) Draw the approximate frequency spectrum of the new digital signal obtained after sampling, and label important points on the axes. (b) What is the bandwidth of the new digital signal obtained after sampling this analog signal? Explain.

3.11 In order to minimize aliasing, what sampling rate should be used to sample a 400-Hz triangular wave? Explain.

3.12 A 100-Hz full-wave-rectified sine wave is sampled at 200 samples/s. The samples are used to directly reconstruct the waveform using a digital-to-analog converter. Will the resulting waveform be a good representation of the original signal? Explain.

3.13 An A/D converter has an input signal range of 10 V. What is the minimum signal that it can resolve (in mV) if it is (a) a 10-bit converter, (b) an 11-bit converter?

3.14 A 10-bit analog-to-digital converter can resolve a minimum signal level of 10 mV. What is the approximate full-scale voltage range of this converter (in volts)?

3.15 A 12-bit D/A converter has an output signal range of ±5 V. What is the approximate minimal step size that it produces at its output (in mV)?

3.16 For an analog-to-digital converter with a full-scale input range of +5 V, how many bits are required to assure resolution of 0.5-mV signal levels?

3.17 A normal QRS complex is about 100 ms wide. (a) What is the American Heart Association's (AHA) specified sampling rate for clinical electrocardiography? (b) If you sample an ECG at the AHA standard sampling rate, about how many sampled data points will define a normal QRS complex?

3.18 An ECG with a 1-mV peak-to-peak QRS amplitude is passed through a filter with a very sharp cutoff, 100-Hz passband, and sampled at 200 samples/s. The ECG is immediately reconstructed with a digital-to-analog converter (DAC) followed by a low-pass reconstruction filter. Comparing the DAC output with the original signal, comment on any differences in appearance due to (a) aliasing, (b) the sampling process itself, (c) the peak-to-peak amplitude, and (d) the clinical acceptability of such a signal.

3.19 An ECG with a 1-mV peak-to-peak QRS amplitude and a 100-ms duration is passed through an ideal low-pass filter with a 100-Hz cutoff. The ECG is then sampled at 200 samples/s. Due to a lack of memory, every other data point is thrown away after the sampling process, so that 100 data points per second are stored. The ECG is immediately reconstructed with a digital-to-analog converter followed by a low-pass reconstruction filter. Comparing the reconstruction filter output with the original signal, comment on any differences in appearance due to (a) aliasing, (b) the sampling process itself, (c) the peak-to-peak amplitude, and (d) the clinical acceptability of such a signal.

3.20 An IBM PC signal acquisition board with an 8-bit A/D converter is used to sample an ECG. An ECG amplifier provides a peak-to-peak signal of 1 V centered in the 0-to-5-V input range of the converter. How many bits of the A/D converter are used to represent the signal?

3.21 A commercial 12-bit signal acquisition board with a ±10-V input range is used to sample an ECG. An ECG amplifier provides a peak-to-peak signal of ±1 V. How many discrete amplitude steps are used to represent the ECG signal?

3.22 Explain the relationship between the frequencies present in a signal and the sampling theorem.

3.23 Describe the effects of having a nonideal (a) input filter; (b) output filter.

3.24 What sampling rate and filter characteristics (e.g., cutoff frequency) would you use to sample an ECG signal?

3.25 What type of A/D converter circuit provides the fastest sampling speed?

3.26 What type of A/D converter circuit tends to average out high-frequency noise?

3.27 In an 8-bit successive-approximation A/D converter, what is the initial digital approximation to a signal?

3.28 A 4-bit successive-approximation A/D converter gets a final approximation to a signal of 0110. What approximation did it make just prior to this final result?

3.29 In a D/A converter design, what are the advantages of an R-$2R$ resistor network over a binary-weighted resistor network?

3.30 For an 8-bit successive approximation analog-to-digital converter, what will be the next approximation made by the converter (in hexadecimal) if the approximation of (a) 0x90 to the input signal is found to be too low, (b) 0x80 to the input signal is found to be too high?

3.31 For an 8-bit successive-approximation analog-to-digital converter, what are the possible results of the next approximation step (in hexadecimal) if the approximation at a certain step is a) 0x10, b) 0x20?

3.32 An 8-bit analog-to-digital converter has a clock that drives the internal successive approximation circuitry at 80 kHz. (a) What is the fastest possible sampling rate that could be achieved by this converter? (b) If 0x80 represents a signal level of 1 V, what is the minimum signal that this converter can resolve (in mV)?

3.33 What circuit is used in a signal conversion system to store analog voltage levels? Draw a schematic of such a circuit and explain how it works.

3.34 The internal IBM PC signal acquisition board described in Appendix A is used to sample an ECG. An amplifier amplifies the ECG so that a 1-mV level uses all 12 bits of the converter. What is the smallest ECG amplitude that can be resolved (in μV)?

3.35 An 8-bit successive-approximation analog-to-digital converter is used to sample an ECG. An amplifier amplifies the ECG so that a 1-mV level uses all 8 bits of the converter. What is the smallest ECG amplitude that can be resolved (in μV)?

3.36 The Computers of Wisconsin (COW) A/D converter chip made with CMOS technology includes an 8-bit successive-approximation converter with a 100-μs sampling period. An on-chip analog multiplexer provides for sampling up to 8 channels. (a) With this COW chip, how fast could you sample a single channel (in samples per s)? (b) How fast could you sample each channel if you wanted to use all eight channels? (c) What is the minimal external clock frequency necessary to drive the successive-approximation circuitry for the maximal sampling rate? (d) List two advantages that this chip has over an equivalent one made with TTL technology.

4

Basics of Digital Filtering

Willis J. Tompkins and Pradeep Tagare

In this chapter we introduce the concept of digital filtering and look at the advantages, disadvantages, and differences between analog and digital filters. Digital filters are the discrete domain counterparts of analog filters. Implementation of different types of digital filters is covered in later chapters. There are many good books that expand on the general topic of digital filtering (Antoniou, 1979; Bogner and Constantinides, 1985; Gold and Rader, 1969; Rabiner and Rader, 1972; Stearns, 1975).

4.1 DIGITAL FILTERS

The function of a digital filter is the same as its analog counterpart, but its implementation is very different. Analog filters are implemented using either active or passive electronic circuits, and they operate on continuous waveforms. Digital filters, on the other hand, are implemented using either a digital logic circuit or a computer program and they operate on a sequence of numbers that are obtained by sampling the continuous waveform. The use of digital filters is widespread today because of the easy availability of computers. A computer program can be written to implement almost any kind of digital filter.

There are several advantages of digital filters over analog filters. A digital filter is highly immune to noise because of the way it is implemented (software/digital circuits). Accuracy is dependent only on round-off error, which is directly determined by the number of bits that the designer chooses for representing the variables in the filter. Also it is generally easy and inexpensive to change a filter's operating characteristics (e.g., cutoff frequency). Unlike an analog filter, performance is not a function of factors such as component aging, temperature variation, and power supply voltage. This characteristic is important in medical applications where most of the signals have low frequencies that might be distorted due to the drift in an analog circuit.

4.2 THE z TRANSFORM

As discussed in Chapter 3, the sampling process reduces a continuous signal to a sequence of numbers. Figure 4.1 is a representation of this process which yields the sequence

$$\{a(0), a(T), a(2T), a(3T), \ldots, a(kT)\} \tag{4.1}$$

This set of numbers summarizes the samples of the waveform $a(t)$ at times $0, T, \ldots,$ kT, where T is the sampling period. _(constant)_

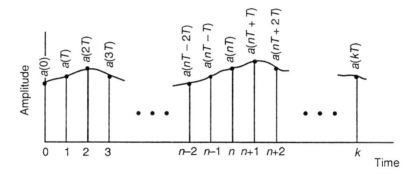

Figure 4.1 Sampling a continuous signal produces a sequence of numbers. Each number is separated from the next in time by the sampling period of T seconds.

We begin our study of digital filters with the z transform. By definition, the z transform of any sequence

$$\{f(0), f(T), f(2T), \ldots, f(kT)\} \tag{4.2}$$

is

$$F(z) = f(0) + f(T)z^{-1} + f(2T)z^{-2} + \ldots + f(kT)z^{-k} \tag{4.3}$$

In general

$$\boxed{F(z) = \sum_{n=0}^{k} f(nT)z^{-n}} \tag{4.4}$$

If we accept this definition, then the z transform of the sequence of Eq. (4.1) is

$$A(z) = a(0) + a(T)z^{-1} + a(2T)z^{-2} + \ldots + a(kT)z^{-k} \tag{4.5}$$

or

$$A(z) = \sum_{n=0}^{k} a(nT)z^{-n} \qquad (4.6)$$

Suppose we want to find the z transform of a signal represented by any sequence, for example

$$\{1, 2, 5, 3, 0, 0, \ldots\} \qquad (4.7)$$

From Eq. (4.4), we can write the z transform at once as

$$X(z) = 1 + 2z^{-1} + 5z^{-2} + 3z^{-3} \qquad (4.8)$$

Since the sequence represents an array of numbers, each separated from the next by the sampling period, we see that the variable z^{-1} in the z transform represents a T-s time separation of one term from the next. The numerical value of the negative exponent of z tells us how many sample periods after the beginning of the sampling process when the sampled data point, which is the multiplier of the z term, occurred. Thus by inspection, we know, for example, that the first sampled data value, which was obtained at $t = 0$, is 1, and that the sample at clock tick 2 (i.e., $t = 2 \times T$ s) is 5.

The z transform is important in digital filtering because it describes the sampling process and plays a role in the digital domain similar to that of the Laplace transform in analog filtering. Figure 4.2 shows two examples of discrete-time signals analogous to common continuous time signals. The unit impulse of Figure 4.2(a), which is analogous to a Dirac delta function, is described by

$$\begin{aligned} f(nT) &= 1 && \text{for } n = 0 \\ f(nT) &= 0 && \text{for } n > 0 \end{aligned} \qquad (4.9)$$

This corresponds to a sequence of

$$\{1, 0, 0, 0, 0, 0, \ldots\} \qquad (4.10)$$

Therefore, the z transform of the unit impulse function is

$$F(z) = 1 \qquad (4.11)$$

This is an important finding since we frequently use the unit impulse as the input function to study the performance of a filter.

For the unit step function in Figure 4.2(b)

$$f(nT) = 1 \qquad \text{for } n \geq 0$$

giving a sequence of

$$\{1, 1, 1, 1, 1, 1, \ldots\} \tag{4.12}$$

Therefore, the z transform of the unit step is

$$F(z) = 1 + z^{-1} + z^{-2} + z^{-3} + \ldots \tag{4.13}$$

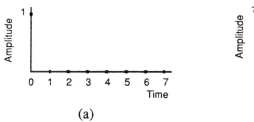 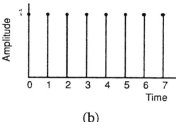

(a)   (b)

Figure 4.2 Examples of discrete-time signals. Variable n is an integer. a) Unit impulse. $\delta(nT) = 1$ for $n = 0$; $\delta(nT) = 0$ for $n \neq 0$. b) Unit step. $f(nT) = 1$ for all n.

This transform is an infinite summation of nonzero terms. We can convert this sum to a more convenient ratio of polynomials by using the binomial theorem

$$1 + v + v^2 + v^3 + \ldots = \frac{1}{1 - v} \tag{4.14}$$

If we let $v = z^{-1}$ in the above equation, the z transform of the unit step becomes

$$F(z) = \frac{1}{1 - z^{-1}} = \frac{z}{z - 1} \tag{4.15}$$

Figure 4.3 summarizes the z transforms of some common signals.

4.3 ELEMENTS OF A DIGITAL FILTER

We need only three types of operations to implement any digital filter: (1) storage for an interval of time, (2) multiplication by a constant, and (3) addition. Figure 4.4 shows the symbols used to represent these operations. Consider the sequence

$$\{x(0), x(T), x(2T), \ldots, x(nT)\} \tag{4.16}$$

$f(t), t \geq 0$	$f(nT), nT \geq 0$	$F(z)$
1(unit step)	1	$\dfrac{1}{1 - z^{-1}}$
t	nT	$\dfrac{Tz^{-1}}{(1 - z^{-1})^2}$
e^{-at}	e^{-anT}	$\dfrac{1}{1 - e^{-aT}z^{-1}}$
te^{-at}	nTe^{-anT}	$\dfrac{Te^{-aT}z^{-1}}{(1 - e^{-aT}z^{-1})^2}$
$\sin\omega_c t$	$\sin n\omega_c T$	$\dfrac{(\sin\omega_c T)z^{-1}}{1 - 2(\cos\omega_c T)z^{-1} + z^{-2}}$
$\cos\omega_c t$	$\cos n\omega_c T$	$\dfrac{1 - (\cos\omega_c T)z^{-1}}{1 - 2(\cos\omega_c T)z^{-1} + z^{-2}}$

Figure 4.3 Examples of continuous time functions $f(t)$ and analogous discrete-time functions $f(nT)$ together with their z transforms.

Its z transform is

$$X(z) = x(0) + x(T)z^{-1} + x(2T)z^{-2} + \ldots + x(nT)z^{-n} \qquad (4.17)$$

If we apply this sequence to the input of the storage element of Figure 4.4(a), we obtain at the output the sequence

$$\{0, x(0), x(T), x(2T), \ldots, x(nT)\} \qquad (4.18)$$

This sequence has the z transform

$$Y(z) = 0 + x(0)z^{-1} + x(T)z^{-2} + \ldots + x(nT - T)z^{-n} \qquad (4.19)$$

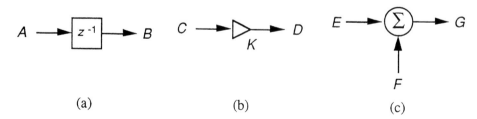

(a)   (b)   (c)

Figure 4.4 Digital filter operators. (a) Storage of a number for one clock period. $B = A$ exactly T seconds after the signal enters A. (b) Multiplication by a constant. $D = K \times C$ instantaneously. (c) Addition of two numbers. $G = E + F$ instantaneously.

In this case, $x(0)$ enters the storage at time $t = 0$; simultaneously, the contents of the block, which is always initialized to 0, are forced to the output. At time $t = T$, $x(0)$ is forced out as $x(T)$ enters. Thus at each clock tick of the analog-to-digital converter from which we get the sampled data, a new number enters the storage block and forces out the previous number (which has been stored for T s). Therefore, the output sequence is identical to the input sequence except that the whole sequence has been delayed by T seconds.

By dividing the output given by Eq. (4.19) by the input of Eq. (4.17), we verify that the relation between the output z transform and the input z transform is

$$Y(z) = X(z)\, z^{-1} \tag{4.20}$$

and the transfer function of the delay block is

$$H(z) = \frac{Y(z)}{X(z)} = z^{-1} \tag{4.21}$$

A microcomputer can easily perform the function of a storage block by placing successive data points in memory for later recall at appropriate clock times.

Figure 4.4(b) shows the second function necessary for implementing a digital filter—multiplication by a constant. For each number in the sequence of numbers that appears at the input of the multiplier, the product of that number and the constant appears instantaneously at the output

$$Y(z) = \beta \times X(z) \tag{4.22}$$

where β is a constant. Ideally there is no storage or time delay in the multiplier. Multiplication of the sequence {5, 9, 0, 6} by the constant 5 would produce a new sequence at the output of the multiplier {25, 45, 0, 30} where each number in the output sequence is said to occur exactly at the same point in time as the corresponding number in the input sequence.

The final operation necessary to implement a digital filter is addition as shown in Figure 4.4(c). At a clock tick, the numbers from two different sequences are summed to produce an output number instantaneously. Again we assume zero delay time for the ideal case.

The relation between the output and the input z transforms is

$$Y(z) = X_1(z) + X_2(z) \tag{4.23}$$

Of course, multiplier and adder circuits require some finite length of time to produce their results, but the delays in a digital filter that control the timing are the storage elements that change their outputs at the rate of the sampling process. Thus, as long as all the arithmetic operations in a digital filter occur within T s, the individual multiply and add operations can be thought of as occurring in zero time, and the filter can perform real-time processing. The nonzero delay time of the adders and multipliers is not a big drawback because in a practical system the outputs from the adders and the multipliers are not required immediately. Since their inputs are constant for almost T s, they actually have T s to generate their outputs.

All general-purpose microprocessors can implement the three operations necessary for digital filtering: storage for a fixed time interval, multiplication, and addition. Therefore they all can implement basic digital filtering. If they are fast enough to do all operations to produce an output value before the next input value appears, they operate in real time. Thus, real-time filters act very much like analog filters in that the filtered signal is produced at the output at the same time as the signal is being applied to the input (with some small delay for processing time).

4.4 TYPES OF DIGITAL FILTERS

The transfer function of a digital filter is the z transform of the output sequence divided by the z transform of the input sequence

$$H(z) = \frac{Y(z)}{X(z)}$$

(4.24)

There are two basic types of digital filters—nonrecursive and recursive. For nonrecursive filters, the transfer function contains a finite number of elements and is in the form of a polynomial

$$H(z) = \sum_{i=0}^{n} h_i z^{-i} = h_0 + h_1 z^{-1} + h_2 z^{-2} + \dots + h_n z^{-n}$$

(4.25)

For recursive filters, the transfer function is expressed as the ratio of two such polynomials

$$H(z) = \frac{\sum_{i=0}^{n} a_i z^{-i}}{1 - \sum_{i=1}^{n} b_i z^{-i}} = \frac{a_0 + a_1 z^{-1} + a_2 z^{-2} + \dots + a_n z^{-n}}{1 - b_1 z^{-1} - b_2 z^{-2} - \dots - b_n z^{-n}}$$

(4.26)

The values of z for which $H(z)$ equals zero are called the zeros of the transfer function, and the values of z for which $H(z)$ goes to infinity are called the poles. We find the zeros of a filter by equating the numerator to 0 and evaluating for z. To find the poles of a filter, we equate the denominator to 0 and evaluate for z. Thus, we can see that the transfer function (and hence the output) goes to zero at the zeros of the transfer function and becomes indeterminate at the poles of the transfer function.

We can see from the transfer functions of nonrecursive filters that they have poles only at $z = 0$. We will see later in this chapter that the location of the poles in the z plane determines the stability of the filter. Since nonrecursive filters have poles only at $z = 0$, they are always stable.

4.5 TRANSFER FUNCTION OF A DIFFERENCE EQUATION

Once we have the difference equation representing the numerical algorithm for implementing a digital filter, we can quickly determine the transfer equation that totally characterizes the performance of the filter. Consider the difference equation

$$y(nT) = x(nT) + 2x(nT - T) + x(nT - 2T) \tag{4.27}$$

Recognizing that $x(nT)$ and $y(nT)$ are points in the input and output sequences associated with the current sample time, they are analogous to the undelayed z-domain variables, $X(z)$ and $Y(z)$ respectively. Similarly $x(nT - T)$, the input value one sample point in the past, is analogous to the z-domain input variable delayed by one sample point, or $X(z)z^{-1}$. We can then write an equation for output $Y(z)$ as a function of input $X(z)$

$$Y(z) = X(z) + 2X(z)z^{-1} + X(z)z^{-2} \tag{4.28}$$

Thus the transfer function of this difference equation is

$$H(z) = \frac{Y(z)}{X(z)} = 1 + 2z^{-1} + z^{-2} \tag{4.29}$$

From this observation of the relationship between discrete-time variables and z-domain variables, we can quickly write the transfer function if we know the difference equation and vice versa.

4.6 THE z-PLANE POLE-ZERO PLOT

We have looked at the z transform from a *practical* point of view. However the mathematics of the z transform are based on the definition

$$\boxed{z = e^{sT}} \tag{4.30}$$

where the complex frequency is

$$s = \sigma + j\omega \tag{4.31}$$

Therefore

$$z = e^{\sigma T} e^{j\omega T} \tag{4.32}$$

By definition, the magnitude of z is

$$|z| = e^{\sigma T} \tag{4.33}$$

and the phase angle is

$$\angle z = \omega T \tag{4.34}$$

If we set $\sigma = 0$, the magnitude of z is 1 and we have

$$z = e^{j\omega T} = \cos\omega T + j\sin\omega T \tag{4.35}$$

This is the equation of a circle of unity radius called the unit circle in the z plane. Since the z plane is a direct mathematical mapping of the well-known s plane, let us consider the stability of filters by mapping conditions from the s domain to the z domain. We will use our knowledge of the conditions for stability in the s domain to study stability conditions in the z domain.

Mapping the s plane to the z plane shows that the imaginary axis $(j\omega)$ in the s plane maps to points on the unit circle in the z plane. Negative values of σ describe the left half of the s plane and map to the interior of the unit circle in the z plane. Positive values of σ correspond to the right half of the s plane and map to points outside the unit circle in the z plane.

For the s plane, poles to the right of the imaginary axis lead to instability. Also any poles on the imaginary axis must be simple. From our knowledge of the mapping between the s and z planes, we can now state the general rule for stability in the z plane. All poles must lie either inside or on the unit circle. If they are on the unit circle, they must be simple. Zeros do not influence stability and can be anywhere in the z plane. Figure 4.5 shows some of the important features of the z plane. Any angle ωT specifies a point on the unit circle. Since $\omega = 2\pi f$ and $T = 1/f_s$, this angle is

$$\omega T = 2\pi \frac{f}{f_s} \tag{4.36}$$

The angular location of any point on the unit circle is then designated by the ratio of a specified frequency f to the sampling frequency f_s. If $f = f_s$, $\omega T = 2\pi$; thus the sampling frequency corresponds to an angular location of 2π radians. For $f = 0$, $\omega T = 0$; hence, dc is located at an angle of $0°$.

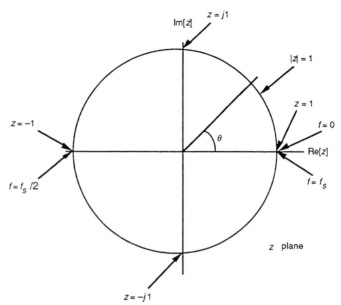

Figure 4.5 Unit circle in the z plane. $\theta = \omega T \approx f \approx f/f_s$, indicating different ways of identifying an angle in the z plane.

Another important frequency is $f = f_s/2 = f_0$ at $\omega T = \pi$. This frequency, called the foldover frequency, equals one-half the sampling rate. Since sampling theory requires a sample rate of twice the highest frequency present in a signal, the foldover frequency represents the maximum frequency that a digital filter can process properly (see Chapter 3). Thus, unlike the frequency axis for the continuous world which extends linearly to infinite frequency, the meaningful frequency axis in the discrete world extends only from 0 to π radians corresponding to a frequency range of dc to $f_s/2$. This is a direct result of the original definition for z in Eq. (4.30) which does a nonlinear mapping of all the points in the s plane into the z plane.

It is important to realize that antialiasing cannot be accomplished with any digital filter except by raising the sampling rate to twice the highest frequency present. This is frequently not practical; therefore, most digital signal processors have an analog front end—the antialias filter.

Figure 4.6 shows that we can refer to the angle designating a point on the unit circle in a number of ways. If we use the ratio f/f_s, it is called the normalized frequency. If we have already established the sampling frequency, we can alternately specify the angular location of a point on the unit circle by frequency f. This illustrates an important feature of a digital filter—the frequency response characteristics are directly related to the sampling frequency. Thus, suppose that the sampling frequency is 200 Hz and a filter has a zero located at 90° on the unit circle (i.e., at 50 Hz). Then if we desire to have the zero of that filter at 25 Hz, a simple way of doing it would be to halve the sampling frequency.

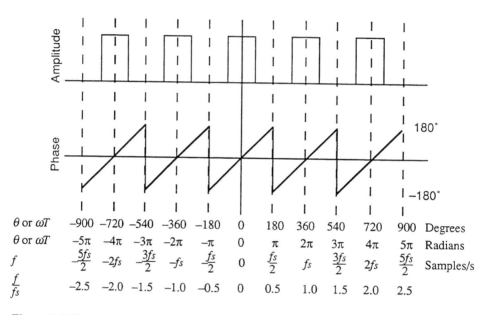

θ or ωT	-900	-720	-540	-360	-180	0	180	360	540	720	900	Degrees
θ or ωT	-5π	-4π	-3π	-2π	$-\pi$	0	π	2π	3π	4π	5π	Radians
f	$-\frac{5f_s}{2}$	$-2f_s$	$-\frac{3f_s}{2}$	$-f_s$	$\frac{f_s}{2}$	0	$\frac{f_s}{2}$	f_s	$\frac{3f_s}{2}$	$2f_s$	$\frac{5f_s}{2}$	Samples/s
$\frac{f}{f_s}$	-2.5	-2.0	-1.5	-1.0	-0.5	0	0.5	1.0	1.5	2.0	2.5	

Figure 4.6 The frequency axis of amplitude and phase responses for digital filters can be labeled in several different ways—as angles, fraction of sampling frequency, or ratio of frequency to sampling frequency. An angle of 360° representing one rotation around the unit circle corresponds to the sampling frequency. The only important range of the amplitude and phase response plots is from 0 to 180° since we restrict the input frequencies to half the sampling frequency in order to avoid aliasing.

As an example, let us consider the following transfer function:

$$H(z) = \frac{1}{3}(1 + z^{-1} + z^{-2}) \tag{4.37}$$

In order to find the locations of the poles and zeros in the z plane, we first multiply by z^2/z^2 in order to make all the exponents of variable z positive.

$$H(z) = \frac{1}{3}(1 + z^{-1} + z^{-2}) \times \frac{z^2}{z^2} = \frac{1}{3}\frac{(z^2 + z + 1)}{z^2} \tag{4.38}$$

Solving for the zeros by setting the numerator equal to zero

$$z^2 + z + 1 = 0 \tag{4.39a}$$

We find that there are two complex conjugate zeros located at

$$z = -0.5 \pm j0.866 \qquad (4.39\text{b})$$

The two zeros are located on the unit circle at $\omega T = \pm 2\pi/3$ ($\pm 120°$). If the sampling frequency is 180 Hz, the zero at $+120°$ will completely eliminate any signal at 60 Hz. Solving for the poles by setting the denominator equal to zero

$$z^2 = 0 \qquad (4.39\text{c})$$

We find that there are two poles, both located at the origin of the z plane

$$z = 0 \qquad (4.39\text{d})$$

These poles are equally distant from all points on the unit circle, so they influence the amplitude response by an equal amount at all frequencies. Because of this, we frequently do not bother to show them on pole-zero plots.

4.7 THE RUBBER MEMBRANE CONCEPT

To get a practical feeling of the pole-zero concept, imagine the z plane to be an infinitely large, flat rubber membrane "nailed" down around its edges at infinity but unconstrained elsewhere. The pole-zero pattern of a filter determines the points at which the membrane is pushed up to infinity (i.e., poles) or is "nailed" to the ground (i.e., zeros). Figure 4.7(a) shows this z-plane rubber membrane with the unit circle superimposed. The magnitude of the transfer function $|H(z)|$ is plotted orthogonal to the x and y axes.

Consider the following transfer function, which represents a pole located in the z plane at the origin, $z = 0$

$$H(z) = z^{-1} = \frac{1}{z} \qquad (4.40)$$

Imagine making a tent by placing an infinitely long tent pole under the rubber membrane located at $z = 0$. Figure 4.7(b) illustrates how this pole appears in a pole-zero plot. Figure 4.7(c) is a view from high above the z plane that shows how the pole stretches the membrane. In Figure 4.7(d) we move our observation location from a distant point high above the rubber membrane to a point just above the membrane since we are principally interested in the influence of the pole directly on the unit circle. Notice how the membrane is distorted symmetrically in all directions from the location of the pole at $z = 0$.

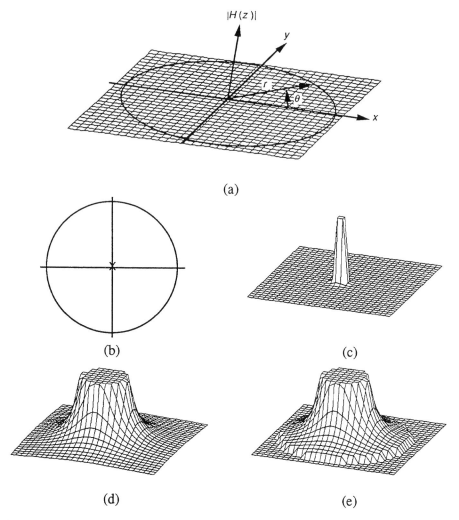

Figure 4.7 Rubber membrane analogy for the z plane. (a) Region of complex z plane showing unit circle, x and y axes, and $|H(z)|$ axis. (b) Pole-zero plot showing single pole at $z = 0$. (c) View of distortion caused by the pole from high above the membrane. (d) View of distortion from viewpoint close to membrane. (e) View with $|H(z)|$ set equal to zero outside the unit circle to visualize how the membrane is stretched on the unit circle itself.

Thus the unit circle is lifted an equal distance from the surface all the way around its periphery. Since the unit circle represents the frequency axis, the amount of stretch of the membrane directly over the unit circle represents $|H(z)|$, the magnitude of the amplitude response.

In order to characterize the performance of a filter, we are principally interested in observations of the amount of membrane stretch directly above the unit circle, and the changes inside and outside the circle are not of particular importance. Therefore in Figure 7(e), we constrain the magnitude of the transfer function to be zero everywhere outside the unit circle in order to be able to better visualize what happens on the unit circle itself. Now we can easily see that the magnitude of the transfer function is the same all the way around the unit circle. We evaluate the magnitude of the transfer function by substituting $z = e^{j\omega T}$ into the function of Eq. (4.40), giving for the complex frequency response

$$H(z) = e^{-j\omega T} = \cos(\omega T) - j\sin(\omega T) \qquad (4.41)$$

The magnitude of this function is

$$|H(\omega T)| = \sqrt{\cos^2(\omega T) + \sin^2(\omega T)} = 1 \qquad (4.42)$$

and the phase response is

$$\angle H(\omega T) = -\omega T \qquad (4.43)$$

Thus the height of the membrane all the way around the unit circle is unity. The magnitude of this function on the unit circle between the angles of $0°$ and $180°$, corresponding to the frequency range of dc to $f_S/2$ respectively, represents the amplitude response of the single pole at $z = 0$. Thus Eq. (4.42) indicates that a signal of any legal frequency entering the input of this unity-gain filter passes through the filter without modification to its amplitude.

The phase response in Eq. (4.43) tells us that an input signal has a phase delay at the output that is linearly proportional to its frequency. Thus this filter is an all-pass filter, since it passes all frequencies equally well, and it has linear phase delay.

In order to see how multiple poles and zeros distort the rubber membrane, let us consider the following transfer function, which has two zeros and two poles.

$$H(z) = \frac{1 - z^{-2}}{1 - 1.0605z^{-1} + 0.5625z^{-2}} \qquad (4.44)$$

To find the locations of the poles and zeros in the z plane, we first multiply by z^2/z^2 in order to make all the exponents of variable z positive.

$$H(z) = \frac{1 - z^{-2}}{1 - 1.0605z^{-1} + 0.5625z^{-2}} \times \frac{z^2}{z^2} = \frac{z^2 - 1}{z^2 - 1.0605z + 0.5625} \qquad (4.45)$$

In order to find the zeros, we set the numerator to zero. This gives

$$z^2 - 1 = 0 \qquad (4.46)$$

or

$$(z + 1)(z - 1) = 0 \qquad (4.47)$$

Therefore, there are two zeros of the function located at

$$z = \pm 1 \tag{4.48}$$

To find the locations of the poles, we set the denominator equal to zero.

$$z^2 - 1.0605z + 0.5625 = 0 \tag{4.49}$$

Solving for z, we obtain a complex-conjugate pair of poles located at

$$z = 0.53 \pm j0.53 \tag{4.50}$$

Figure 4.8(a) is the pole-zero plot showing the locations of the two zeros and two poles for this transfer function $H(z)$. The zeros "nail" down the membrane at the two locations, $z = \pm 1$ (i.e., radius r and angle θ of $1\angle 0°$ and $-1\angle 180°$ in polar notation). The poles stretch the membrane to infinity at the two points, $z = 0.53 \pm j0.53$ (i.e., $0.75\angle \pm 45°$ in polar notation).

Figure 4.8(b) shows the distortion of the unit circle (i.e., the frequency axis). In this view, the plane of Figure 4.8(a) is rotated clockwise around the vertical axis by an azimuth angle of 20° and is tilted at an angle of elevation angle of 40° where the top view in Figure 4.8(a) is at an angle of 90°. The membrane distortion between the angles of 0 to 180° represents the amplitude response of the filter.

The zero at an angle of 0° completely attenuates a dc input. The zero at an angle of 180° (i.e., $f_s/2$) eliminates the highest possible input frequency that can legally be applied to the filter since sampling theory restricts us from putting any frequency into the filter greater than half the sampling frequency. You can see that the unit circle is "nailed" down at these two points. The poles distort the unit circle so as to provide a passband between dc and the highest input frequency at a frequency of $f_s/8$. Thus, this filter acts as a bandpass filter.

The positive frequency axis, which is the top half of the unit circle (see Figure 4.5), is actually hidden from our view in this presentation since it is in the background as we view the rubber membrane with the current viewing angle. Of course, it is symmetrical with the negative frequency axis between angles 0 to $-180°$, which we see in the foreground.

In order to better visualize the amplitude response, we can "walk" around the membrane and look at the hidden positive frequency side. In Figures 4.8(c) to 4.8(f), we rotate clockwise to azimuth angles of 60°, 100°, 140°, and finally 180° respectively. Thus in Figure 4.8(f), the positive frequency axis runs from dc at the left of the image to $f_s/2$ at the right.

We next tilt the image in Figures 4.8(g) and 4.8(h) to elevation angles of 20° and 0° respectively. Thus Figure 4.8(h) provides a direct side view with the horizontal axis representing the edge of the z plane. This view shows us the amplitude response of this filter with a peak output at a frequency corresponding to the location of the pole at an angle of 45° corresponding to a frequency of $f_s/8$.

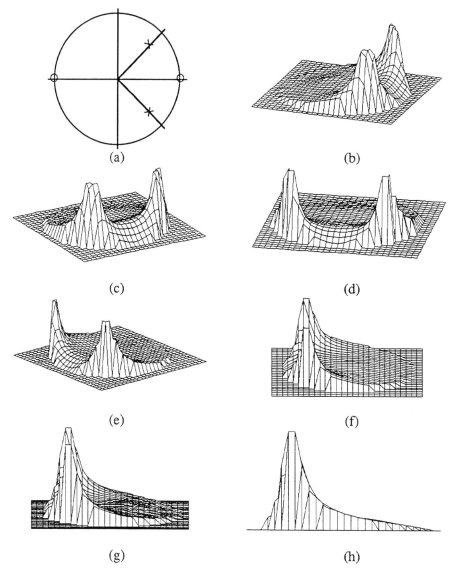

Figure 4.8 A filter with two zeros and two poles. Zeros are located at $z = \pm 1$ and poles at $z = 0.53 \pm j0.53$ (i.e., $r = 0.75$, $\theta = 45°$). Zeros nail the membrane down where they are located. Poles stretch it to infinite heights at the points where they located. (a) Pole-zero plot. Azimuth (AZ) is 0°; elevation (EL) is 90°. (b) Rubber membrane view. AZ = 20°; EL = 40°. (c) AZ = 60°, EL = 40°. (d) AZ = 100°, EL = 40°. (e) AZ = 140°, EL = 40°. (f) AZ = 180°, EL = 40°. (g) AZ = 180°, EL = 20°. (h) AZ = 180°, EL = 0°.

The amplitude reponse clearly goes to zero at the left and right sides of the plot corresponding to the locations of zeros at dc and $f_s/2$. This plot is actually a projection of the response from the circular unit circle axis using a linear amplitude scale.

In order to see how the frequency response looks on a traditional response plot, we calculate the amplitude response by substituting into the transfer function of Eq. (4.44) the equation $z = e^{j\omega T}$

$$H(\omega T) = \frac{1 - e^{-j2\omega T}}{1 - 1.0605e^{-j\omega T} + 0.5625e^{-j2\omega T}} \qquad (4.51)$$

We now substitute into this function the relationship

$$e^{j\omega T} = \cos(\omega T) + j\sin(\omega T) \qquad (4.52)$$

giving

$$H(\omega T) = \qquad\qquad (4.53)$$
$$\frac{1 - \cos(2\omega T) + j\sin(2\omega T)}{1 - 1.0605\cos(\omega T) + j1.0605\sin(\omega T) + 0.5625\cos(2\omega T) - j0.5625\sin(2\omega T)}$$

Collecting real and imaginary terms gives

$$H(\omega T) = \qquad\qquad (4.54)$$
$$\frac{[1 - \cos(2\omega T)] + j[\sin(2\omega T)]}{[1 - 1.0605\cos(\omega T) + 0.5625\cos(2\omega T)] + j[1.0605\sin(\omega T) - 0.5625\sin(2\omega T)]}$$

This equation has the form

$$H(\omega T) = \frac{A + jB}{C + jD} \qquad (4.55)$$

In order to find the amplitude and phase responses of this filter, we first multiply the numerator and denominator by the complex conjugate of the denominator

$$H(\omega T) = \frac{A + jB}{C + jD} \times \frac{C - jD}{C - jD} = \frac{(AC + BD) + j(BC - AD)}{C^2 + D^2} \qquad (4.56)$$

The amplitude response is then

$$|H(\omega T)| = \frac{\sqrt{((AC + BD)^2 + (BC - AD)^2)}}{C^2 + D^2} \qquad (4.57)$$

This response curve is plotted in Figure 4.9(a). Note that the abscissa is f/f_s, so that the range of the axis goes from 0 (dc) to 0.5 ($f_s/2$).

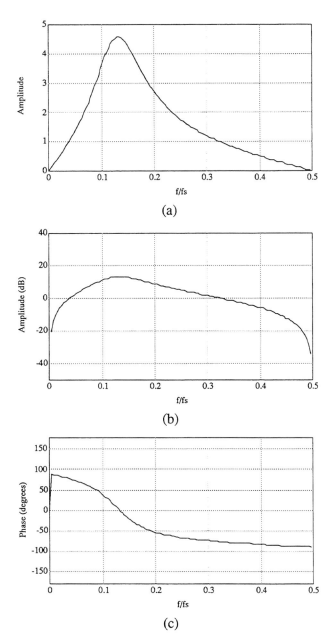

Figure 4.9 Frequency response for the filter of Figure 4.8. (a) Amplitude response with linear amplitude scale. (b) Amplitude response with decibel amplitude scale. (c) Phase response.

Compare this response to Figure 4.8(h) which shows the same information as a projection of the response from the circular unit circle axis. Figure 4.9(b) shows the same amplitude response plotted as a more familiar decibel (dB) plot.

Figure 4.10 shows how the radial locations of the poles influence the distortion of the rubber membrane. Note how moving the poles toward the unit circle makes the peaks appear sharper. The closer the poles get to the unit circle, the sharper the rolloff of the filter.

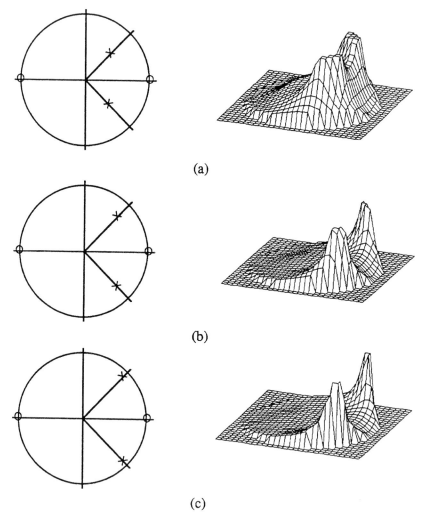

(a)

(b)

(c)

Figure 4.10 Pole-zero and rubber membrane plots of bandpass filter. Angular frequency of pole locations is $f_s/8$. (a) r = 0.5. (b) r = 0.75. (c) r = 0.9.

The phase response for this filter, which is plotted in Figure 4.9(c), is

$$\angle H(\omega T) = \tan^{-1}\left(\frac{BC - AD}{AC + BD}\right) \qquad (4.58)$$

Figure 4.11(a) gives a superposition of the amplitude responses for each of these three pole placements. Note how the filter with the pole closest to the unit circle has the sharpest rolloff (i.e., the highest Q) of the three filters, thereby providing the best rejection of frequencies outside its 3-dB passband.

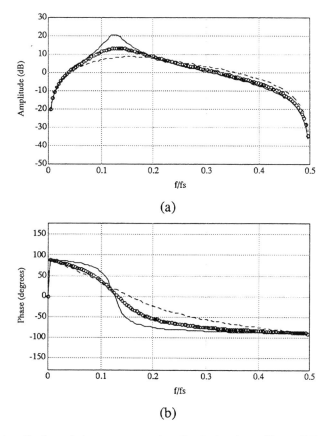

(a)

(b)

Figure 4.11 Amplitude and phase response plots for the bandpass filters of Figure 4.10, all with poles at angles of ±45° (i.e., $f/f_s = 0.125$). Solid line: Poles at $r = 0.9$. Circles: $r = 0.75$. Dashed line: $r = 0.5$.

As the pole moves toward the center of the unit circle, the rolloff becomes decreasingly sharp. Figure 4.11(b) shows the phase responses of these three filters. Note how the phase response becomes more and more linear as the poles are moved toward the center of the unit circle. We will see in the next chapter that the phase response becomes completely piecewise linear when all the poles are at the center of the unit circle (i.e., at $r = 0$).

Thus, the rubber membrane concept helps us to visualize the behavior of any digital filter and can be used as an important tool in the design of digital filters.

4.8 REFERENCES

Antoniou, A. 1979. *Digital Filters: Analysis and Design,* New York: McGraw-Hill.
Bogner, R. E. and Constantinides, A. G. 1985. *Introduction to Digital Filtering,* New York: John Wiley and Sons.
Gold, B. and Rader, C. 1969. *Digital Processing of Signals,* New York: Lincoln Laboratory Publications, McGraw-Hill.
Rabiner, L. R. and Rader, C. M. 1972. *Digital Signal Processing* New York: IEEE Press.
Stearns, S. D. 1975. *Digital Signal Analysis*. Rochelle Park, NJ: Hayden.

4.9 STUDY QUESTIONS

4.1   What are the differences between an analog filter and a digital filter?
4.2   If the output sequence of a digital filter is {1, 0, 0, 2, 0, 1} in response to a unit impulse, what is the transfer function of this filter?
4.3   Draw the pole-zero plot of the filter described by the following transfer function:

$$H(z) = \frac{1}{4} + \frac{1}{2}z^{-1} + \frac{1}{4}z^{-2}$$

4.4   Suppose you are given a filter with a zero at $30°$ on the unit circle. You are asked to use this filter as a notch filter to remove 60-Hz noise. How will you do this? Can you use the same filter as a notch filter, rejecting different frequencies?
4.5   What is the z transform of a step function having an amplitude of five {i.e., 5, 5, 5, 5, ...}?
4.6   A function e^{-at} is to be applied to the input of a filter. Derive the z transform of the discrete version of this function.
4.7   Application of a *unit impulse* to the input of a filter whose performance is unknown produces the output sequence {1, –2, 0, 0, ...}. What would the output sequence be if a *unit step* were applied?
4.8   A digital filter has the transfer function: $H(z) = z^{-1} + 6z^{-4} - 2z^{-7}$. What is the difference equation for the output, $y(nT)$?
4.9   A digital filter has the output sequence {1, 2, –3, 0, 0, 0, ...} when its input is the *unit impulse* {1, 0, 0, 0, 0, ...}. If its input is a *unit step*, what is its output sequence?
4.10 A *unit impulse* applied to a digital filter results in the output sequence: {3, 2, 3, 0, 0, 0, ...}. A *unit step* function applied to the input of the same filter would produce what output sequence?
4.11 The z transform of a filter is: $H(z) = 2 - 2z^{-4}$. What is its (a) amplitude response, (b) phase response, (c) difference equation?

4.12 The transfer function of a filter designed for a sampling rate of 800 samples/s is:

$$H(z) = (1 - 0.5z^{-1})(1 + 0.5z^{-1})$$

A 200-Hz sine wave with a peak amplitude of 4 is applied to the input. What is the peak value of the output signal?

4.13 A *unit impulse* applied to a digital filter results in the following output sequence: {1, 2, 3, 4, 0, 0, ...}. A *unit step* function applied to the input of the same filter would produce what output sequence?

4.14 The transfer function of a filter designed for a sampling rate of 600 samples/s is:

$$H(z) = 1 - 2z^{-1}$$

A sinusoidal signal is applied to the input: 10 sin(628t). What is the peak value of the output signal?

5

Finite Impulse Response Filters

Jesse D. Olson

A finite impulse response (FIR) filter has a unit impulse response that has a limited number of terms, as opposed to an infinite impulse response (IIR) filter which produces an infinite number of output terms when a unit impulse is applied to its input. FIR filters are generally realized nonrecursively, which means that there is no feedback involved in computation of the output data. The output of the filter depends only on the present and past inputs. This quality has several important implications for digital filter design and applications. This chapter discusses several FIR filters typically used for real-time ECG processing, and also gives an overview of some general FIR design techniques.

5.1 CHARACTERISTICS OF FIR FILTERS

5.1.1 Finite impulse response

Finite impulse response implies that the effect of transients or initial conditions on the filter output will eventually die away. Figure 5.1 shows a signal-flow graph (SFG) of a FIR filter realized nonrecursively. The filter is merely a set of "tap weights" of the delay stages. The unit impulse response is equal to the tap weights, so the filter has a difference equation given by Eq. (5.1), and a transfer function equation given by Eq. (5.2).

$$y(nT) = \sum_{k=0}^{N} b_k \, x(nT - kT) \qquad (5.1)$$

$$H(z) = b_0 + b_1 z^{-1} + b_2 z^{-2} + \ldots + b_N z^{-N} \qquad (5.2)$$

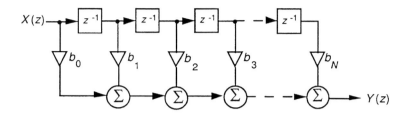

Figure 5.1 The output of a FIR filter of order N is the weighted sum of the values in the storage registers of the delay line.

5.1.2 Linear phase

In many biomedical signal processing applications, it is important to preserve certain characteristics of a signal throughout the filtering operation, such as the height and duration of the QRS pulse. A filter with linear phase has a pure time delay as its phase response, so phase distortion is minimized. A filter has linear phase if its frequency response $H(e^{j\theta})$ can be expressed as

$$H(e^{j\theta}) = H_1(\theta)\, e^{-j(\alpha\theta + \beta)} \tag{5.3}$$

where $H_1(\theta)$ is a real and even function, since the phase of $H(e^{j\theta})$ is

$$\angle H(e^{j\theta}) = \begin{cases} -\alpha\theta - \beta & ; H_1(\theta) > 0 \\ -\alpha\theta - \beta - \pi & ; H_1(\theta) < 0 \end{cases} \tag{5.4}$$

FIR filters can easily be designed to have a linear phase characteristic. Linear phase can be obtained in four ways, as combinations of even or odd symmetry (defined as follows) with even or odd length.

$$\left. \begin{array}{l} h(N-1-k) = h(k),\ even\ symmetry \\ h(N-1-k) = -h(k),\ odd\ symmetry \end{array} \right\} \quad \text{for} \quad 0 \le k \le N \tag{5.5}$$

5.1.3 Stability

Since a nonrecursive filter does not use feedback, it has no poles except those that are located at $z = 0$. Thus there is no possibility for a pole to exist outside the unit circle. This means that it is inherently stable. As long as the input to the filter is bounded, the output of the filter will also be bounded. This contributes to ease of design, and makes FIR filters especially useful for adaptive filtering where filter coefficients change as a function of the input data. Adaptive filters are discussed in Chapter 8.

5.1.4 Desirable finite-length register effects

When data are converted from analog form to digital form, some information is lost due to the finite number of storage bits. Likewise, when coefficient values for a filter are calculated, digital implementation can only approximate the desired values. The limitations introduced by digital storage are termed finite-length register effects. Although we will not treat this subject in detail in this book, finite-length register effects can have significant negative impact on a filter design. These effects include quantization error, roundoff noise, limit cycles, conditional stability, and coefficient sensitivity. In FIR filters, these effects are much less significant and easier to analyze than in IIR filters since the errors are not fed back into the filter. See Appendix F for more details about finite-length register effects.

5.1.5 Ease of design

All of the above properties contribute to the ease in designing FIR filters. There are many straightforward techniques for designing FIR filters to meet arbitrary frequency and phase response specifications, such as window design or frequency sampling. Many software packages exist that automate the FIR design process, often computing a filter realization that is in some sense optimal.

5.1.6 Realizations

There are three methods of realizing an FIR filter (Bogner and Constantinides, 1985). The most common method is direct convolution, in which the filter's unit impulse sequence is convolved with the present input and past inputs to compute each new output value. FIR filters attenuate the signal very gradually outside the passband (i.e., they have slow rolloff characteristics). Since they have significantly slower rolloff than IIR filters of the same length, for applications that require sharp rolloffs, the order of the FIR filter may be quite large. For higher-order filters the direct convolution method becomes computationally inefficient.

For FIR filters of length greater than about 30, the "fast convolution" realization offers a computational savings. This technique takes advantage of the fact that time-domain multiplication, the frequency-domain dual of convolution, is computationally less intensive. Fast convolution involves taking the FFT of a block of data, multiplying the result by the FFT of the unit impulse sequence, and finally taking the inverse FFT. The process is repeated for subsequent blocks of data. This method is discussed in detail in section 11.3.2.

The third method of realizing FIR filters is an advanced, recursive technique involving a comb filter and a bank of parallel digital resonators (Rabiner and Rader, 1972). This method is advantageous for frequency sampling designs if a large number of the coefficients in the desired frequency response are zero, and can be used for filters with integer-valued coefficients, as discussed in Chapter 7. For the remainder of this chapter, only the direct convolution method will be considered.

5.2 SMOOTHING FILTERS

One of the most common signal processing tasks is smoothing of the data to reduce high-frequency noise. Some sources of high-frequency noise include 60-Hz, movement artifacts, and quantization error. One simple method of reducing high-frequency noise is to simply average several data points together. Such a filter is referred to as a moving average filter.

5.2.1 Hanning filter

One of the simplest smoothing filters is the Hanning moving average filter. Figure 5.2 summarizes the details of this filter. As illustrated by its difference equation, the Hanning filter computes a weighted moving average, since the central data point has twice the weight of the other two:

$$y(nT) = \frac{1}{4}\left[x(nT) + 2x(nT - T) + x(nT - 2T)\right] \tag{5.6}$$

As we saw in section 4.5, once we have the difference equation representing the numerical algorithm for implementing a digital filter, we can quickly determine the transfer equation that totally characterizes the performance of the filter by using the analogy between discrete-time variables and z-domain variables.

Recognizing that $x(nT)$ and $y(nT)$ are points in the input and output sequences associated with the current sample time, they are analogous to the undelayed z-domain variables, $X(z)$ and $Y(z)$ respectively. Similarly $x(nT - T)$, the input value one sample point in the past, is analogous to the z-domain input variable delayed by one sample point, or $X(z)z^{-1}$. We can then write an equation for output $Y(z)$ as a function of input $X(z)$:

$$Y(z) = \frac{1}{4}[X(z) + 2X(z)z^{-1} + X(z)z^{-2}] \tag{5.7}$$

The block diagram of Figure 5.2(a) is drawn using functional blocks to directly implement the terms in this equation. Two delay blocks are required as designated by the -2 exponent of z. Two multipliers are necessary to multiply by the factors 2 and 1/4, two summers are needed to combine the terms. The transfer function of this equation is

$$H(z) = \frac{1}{4}\left[1 + 2z^{-1} + z^{-2}\right] \tag{5.8}$$

This filter has two zeros, both located at $z = -1$, and two poles, both located at $z = 0$ (see section 4.6 to review how to find pole and zero locations). Figure 5.2(b) shows the pole-zero plot. Note the poles are implicit; they are not drawn since they influence all frequencies in the amplitude response equally.

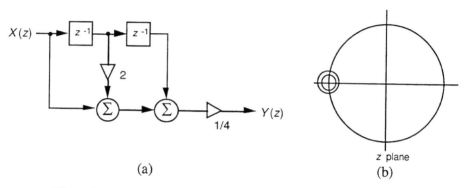

(a)   (b)

Figure 5.2 Hanning filter. (a) Signal-flow graph. (b) Pole-zero diagram.

The filter's amplitude and phase responses are found by substituting $e^{j\omega T}$ for z in Eq. (5.8):

$$H(\omega T) = \frac{1}{4}\left[1 + 2e^{-j\omega T} + e^{-j2\omega T}\right] \qquad (5.9)$$

We could now directly substitute into this function the trigonometric relationship

$$e^{j\omega T} = \cos(\omega T) + j\sin(\omega T) \qquad (5.10)$$

However, a common trick prior to this substitution that leads to quick simplification of expressions such as this one is to extract a power of e as a multiplier such that the final result has two similar exponential terms with equal exponents of opposite sign

$$H(\omega T) = \frac{1}{4}\left[e^{-j\omega T}\left(e^{j\omega T} + 2 + e^{-j\omega T}\right)\right] \qquad (5.11)$$

Now substituting Eq. (5.10) for the terms in parentheses yields

$$H(\omega T) = \frac{1}{4}\left\{e^{-j\omega T}\left[\cos(\omega T) + j\sin(\omega T) + 2 + \cos(\omega T) - j\sin(\omega T)\right]\right\} \qquad (5.12)$$

The $\sin(\omega T)$ terms cancel leaving

$$H(\omega T) = \frac{1}{4} \left[(2 + 2\cos(\omega T))e^{-j\omega T} \right] \qquad (5.13)$$

This is of the form $Re^{j\theta}$ where R is the real part and θ is the phase angle. Thus, the magnitude response of the Hanning filter is $|R|$, or

$$|H(\omega T)| = \left| \frac{1}{2} \left[1 + \cos(\omega T) \right] \right| \qquad (5.14)$$

Figure 5.3(a) shows this cosine-wave amplitude response plotted with a linear ordinate scale while Figure 5.3(b) shows the same response using the more familiar decibel plot, which we will use throughout this book. The relatively slow rolloff of the Hanning filter can be sharpened by passing its output into the input of another identical filter. This process of connecting multiple filters together is called cascading filters. The linear phase response shown in Figure 5.3(c) is equal to angle θ, or

$$\angle H(\omega T) = -\omega T \qquad (5.15)$$

Implementation of the Hanning filter is accomplished by writing a computer program. Figure 5.4 illustrates a C-language program for an off-line (i.e., not real-time) application where data has previously been sampled by an A/D converter and left in an array. This program directly computes the filter's difference equation [Eq. (5.6)]. Within the `for()` loop, a value for $x(nT)$ (called `xnt` in the program) is obtained from the array `idb[]`. The difference equation is computed to find the output value $y(nT)$ (or `ynt` in the program). This value is saved into the data array, replacing the value of $x(nT)$. Then the input data variables are shifted through the delay blocks. Prior to the next input, the data point that was one point in the past $x(nT - T)$ (called `xm1` in the program) moves two points in the past and becomes $x(nT - 2T)$ (or `xm2`). The most recent input $x(nT)$ (called `xnt`) moves one point back in time, replacing $x(nT - T)$ (or `xm1`). In the next iteration of the `for()` loop, a new value of $x(nT)$ is retrieved, and the process repeats until all 256 array values are processed. The filtered output waveform is left in array `idb[]`.

The Hanning filter is particularly efficient for use in real-time applications since all of its coefficients are integers, and binary shifts can be used instead of multiplications. Figure 5.5 is a real-time Hanning filter program. In this program, the computation of the output value $y(nT)$ must be accomplished during one sample interval T. That is, every new input data point acquired by the A/D converter must produce an output value before the next A/D input. Otherwise the filter would not keep up with the sampling rate, and it would not be operating in real time.

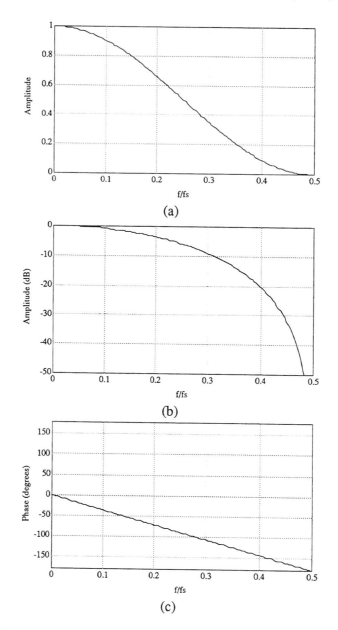

Figure 5.3 Hanning filter. (a) Frequency response (linear magnitude axis). (b) Frequency response (dB magnitude axis). (c)Phase response.

In this program, sampling from the A/D converter, computation of the results, and sending the filtered data to a D/A converter are all accomplished within a `for()` loop. The `wait()` function is designed to wait for an interrupt caused by an A/D clock tick. Once the interrupt occurs, a data point is sampled with the `adget()` function and set equal to `xnt`. The Hanning filter's difference equation is then computed using C-language shift operators to do the multiplications efficiently. Expression `<<1` is a binary shift to the left by one bit position corresponding to multiplication by a factor of two, and `>>2` is a binary shift right of two bit positions representing division by four.

```
/*   Hanning filter
     Difference equation:
         y(nT)  =  (x(nT)  + 2*x(nT - T) + x(nT - 2T))/4
     C language implementation equation:
         ynt  =  (xnt + 2*xm1 + xm2)/4;
*/

main()
{
        int i, xnt, xm1, xm2, ynt, idb[256];

        xm2 = 0;
        xm1 = 0;

        for(i = 0; i <= 255; i++)
            {
            xnt = idb[i];
            ynt = (xnt + 2*xm1 + xm2)/4;
            idb[i] = ynt;
            xm2 = xm1;
            xm1 = xnt;
            }
}
```

Figure 5.4 C-language code to implement the Hanning filter. Data is presampled by an ADC and stored in array `idb[]`. The filtered signal is left in `idb[]`.

The computed output value is sent to a D/A converter with function `daput()`. Then the input data variables are shifted through the delay blocks as in the previous program. For the next input, the data point that was one point in the past $x(nT - T)$ (called `xm1` in the program) moves two points in the past and becomes $x(nT - 2T)$ (or `xm2`). The most recent input $x(nT)$ (called `xnt`) moves one point back in time, replacing $x(nT - T)$ (or `xm1`). Then the `for()` loop repeats with the `wait()` function waiting until the next interrupt signals that a new sampled data point is available to be acquired by `adget()` as the new value for $x(nT)$.

```
/*  Real-time Hanning filter
    Difference equation:
        y(nT) = (x(nT) + 2*x(nT - T) + x(nT - 2T))/4
    C language implementation equation:
        ynt = (xnt + xm1<<1 + xm2)>>2;
*/

#define  AD  31;
#define  DA  32;

main()
{
        int i, xnt, xm1, xm2, ynt;

        xm2 = 0;
        xm1 = 0;

        tmic 2000;          /* Start ADC clock ticking at 2000 µs  */
                            /* intervals (2 ms period for 500 sps) */

        for( ; ; )
            {
            wait();                 /* Wait for ADC clock to tick  */
            xnt = adget(AD);
            ynt = (xnt + xm1<<1 + xm2)>>2;
            daput(ynt, DA);
            xm2 = xm1;
            xm1 = xnt;
            }
}
```

Figure 5.5 C-language code to implement the real-time Hanning filter.

5.2.2 Least-squares polynomial smoothing

This family of filters fits a parabola to an odd number $(2L + 1)$ of input data points in a least-squares sense. Figure 5.6(a) shows that the output of the filter is the midpoint of the parabola. Writing the equation for a parabola at each input point, we obtain

$$p(nT + kT) = a(nT) + b(nT)k + c(nT)k^2 \qquad (5.16)$$

where k ranges from $-L$ to L. The fit is found by selecting $a(nT)$, $b(nT)$ and $c(nT)$ to minimize the squared error between the parabola and the input data. Setting the partial derivatives of the error with respect to $a(nT)$, $b(nT)$, and $c(nT)$ equal to zero results in a set of simultaneous equations in $a(nT)$, $b(nT)$, $c(nT)$, k, and $p(nT - kT)$. Solving to obtain an expression for $a(nT)$, the value of the parabola at $k = 0$, yields an expression that is a function of the input values. The coefficients of this expression are the tap weights for the least-squares polynomial filter as shown in the signal-flow graph of Figure 5.6(b) for a five-point filter. The difference equation for the five-point parabolic filter is

$$y(nT) = \frac{1}{35} \left[(-3x(nT) + 12x(nT - T) + 17x(nT - 2T) \right.$$
$$\left. + 12x(nT - 3T) - 3x(nT - 4T)) \right] \tag{5.17}$$

Its transfer function is

$$H(z) = \frac{1}{35} \left[-3 + 12z^{-1} + 17z^{-2} + 12z^{-3} - 3z^{-4} \right] \tag{5.18}$$

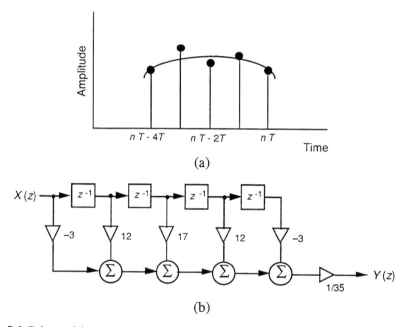

(a)

(b)

Figure 5.6 Polynomial smoothing filter with $L = 2$. (a) Parabolic fitting of groups of 5 sampled data points. (b) Signal-flow graph.

Figure 5.7 shows the tap weights for filters with L equal to 2, 3, 4, and 5, and Figure 5.8 illustrates their responses. The order of the filter can be chosen to meet the desired rolloff.

L	Tap weights
2	$\frac{1}{35}(-3,\ 12,\ 17,\ 12,\ -3)$
3	$\frac{1}{21}(-2,\ 3,\ 6,\ 7,\ 6,\ 3,\ -2)$
4	$\frac{1}{231}(-21,\ 14,\ 39,\ 54,\ 59,\ 54,\ 39,\ 14,\ -21)$
5	$\frac{1}{429}(-36,\ 9,\ 44,\ 69,\ 84,\ 89,84,69,44,\ 9,-36)$

Figure 5.7 Tap weights of polynomial smoothing filters (Hamming, 1977).

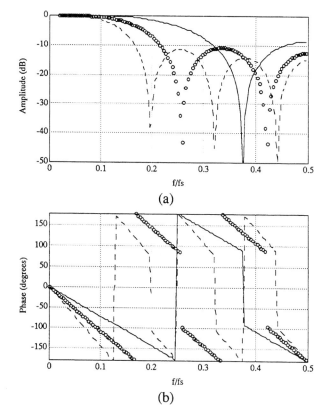

(a)

(b)

Figure 5.8 Polynomial smoothing filters (a) Amplitude responses. (b) Phase responses. Solid line: $L = 2$. Circles: $L = 3$. Dashed line: $L = 4$.

5.3 NOTCH FILTERS

A common biomedical signal processing problem involves the removal of noise of a particular frequency or frequency range (such as 60 Hz) from a signal while passing higher and/or lower frequencies without attenuation. A filter that performs this task is referred to as a notch, bandstop, or band-reject filter.

One simple method of completely removing noise of a specific frequency from the signal is to place a zero on the unit circle at the location corresponding to that frequency. For example, if a sampling rate of 180 samples per second is used, a zero at $2\pi/3$ removes 60-Hz line frequency noise from the signal. The difference equation is

$$y(nT) = \frac{1}{3}\left[x(nT) + x(nT - T) + x(nT - 2T)\right] \qquad (5.19)$$

The filter has zeros at

$$z = -0.5 \pm j\,0.866 \qquad (5.20)$$

and its amplitude and phase response are given by

$$|H(\omega T)| = \left|\frac{1}{3}\left[1 + 2\cos(\omega T)\right]\right| \qquad (5.21)$$

$$\angle H(\omega T) = -\omega T \qquad (5.22)$$

Figure 5.9 shows the details of the design and performance of this filter. The relatively slow rolloff of this filter causes significant attenuation of frequencies other than 60 Hz as well.

5.4 DERIVATIVES

The response of the true derivative function increases linearly with frequency. However, for digital differentiation such a response is not possible since the frequency response is periodic. The methods discussed in this section offer trade-offs between complexity of calculation, approximation to the true derivative, and elimination of high-frequency noise. Figure 5.10 shows the signal-flow graphs and pole-zero plots for three different differentiation algorithms: two-point difference, three-point central difference, and least-squares polynomial fit.

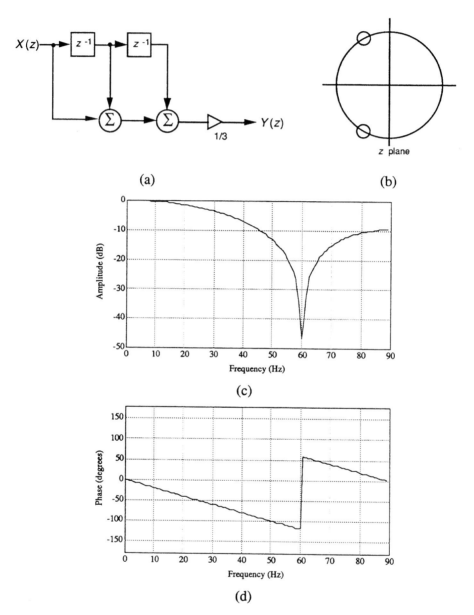

(a)

(b)

(c)

(d)

Figure 5.9 The 60-Hz notch filter. (a) Signal-flow graph. (b) Pole-zero plot. (c) Frequency response. (d) Phase response.

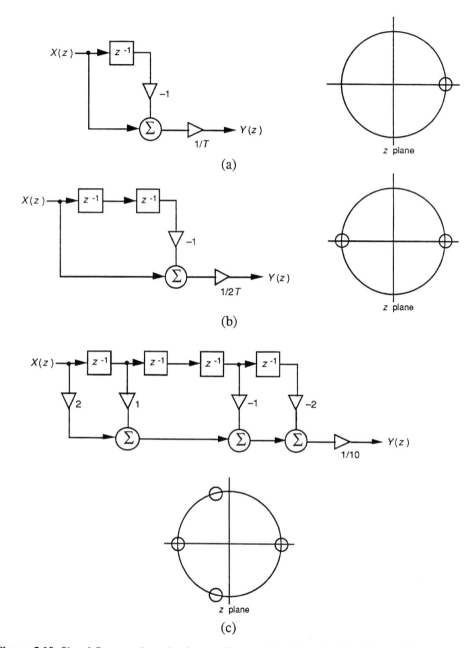

Figure 5.10 Signal flow graphs and pole-zero diagrams for derivative algorithms. (a) Two-point. (b) Three-point central difference. (c) 5-point least squares polynomial.

5.4.1 Two-point difference

The two-point difference algorithm, the simplest of these derivative algorithms, places a zero at $z = 1$ on the unit circle. Its amplitude response shown in Figure 5.11(a) closely approximates the ideal response, but since it does not go to zero at $f_S/2$, it greatly amplifies high-frequency noise. It is often followed by a low-pass filter. Its difference equation is

$$y(nT) = \frac{1}{T}\left[x(nT) - x(nT - T)\right] \qquad (5.23)$$

Its transfer function is

$$H(z) = \frac{1}{T}\left(1 - z^{-1}\right) \qquad (5.24)$$

5.4.2 Three-point central difference

The three-point central difference algorithm places zeros at $z = 1$ and $z = -1$, so the approximation to the derivative is poor above $f_S/10$ seen in Figure 5.11(a). However, since the response goes to zero at $f_S/2$, the filter has some built-in smoothing. Its difference equation is

$$y(nT) = \frac{1}{2T}\left[x(nT) - x(nT - 2T)\right] \qquad (5.25)$$

Its transfer function is

$$H(z) = \frac{1}{2T}\left(1 - z^{-2}\right) \qquad (5.26)$$

5.4.3 Least-squares polynomial derivative approximation

This filter is similar to the parabolic smoothing filter described earlier, except that the slope of the polynomial is taken at the center of the parabola as the value of the derivative. The coefficients of the transfer equations for filters with $L = 2, 3, 4$, and 5 are illustrated in Figure 5.12. Figure 5.10(c) shows the signal-flow graph and pole-zero diagram for this filter with $L = 2$. Note that the filter has zeros at $z = \pm 1$, as did the three-point central difference, with additional zeros at $z = -0.25 \pm j0.968$. The difference equation for the five-point parabolic filter is

$$y(nT) = \frac{1}{10T}\left[2x(nT) + x(nT - T) - x(nT - 3T) - 2x(nT - 4T)\right] \qquad (5.27)$$

Its transfer function is

$$H(z) = \frac{1}{10T} \left[2 + z^{-1} - z^{-3} - 2z^{-4} \right] \tag{5.28}$$

As Figure 5.11(a) shows, the response only approximates the true derivative at low frequencies, since the smoothing nature of the parabolic fit attenuates high frequencies significantly.

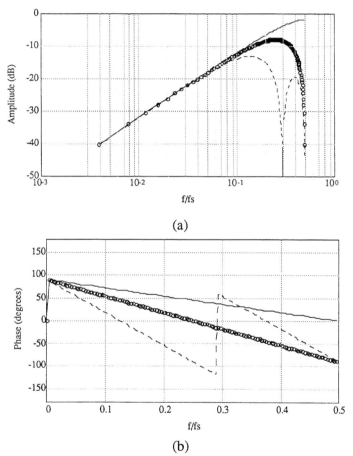

(a)

(b)

Figure 5.11 Derivatives. (a) Amplitude response. (b) Phase response. Solid line: Two-point. Circles: Three-point central difference. Dashed line: Least-squares parabolic approximation for $L = 2$.

L	Tap weights
2	$\frac{1}{10T}\,(2, 1, 0, -1, -2)$
3	$\frac{1}{28T}\,(3, 2, 1, 0, -1, -2, -3)$
4	$\frac{1}{60T}\,(4, 3, 2, 1, 0, -1, -2, -3, -4)$
5	$\frac{1}{110T}\,(5, 4, 3, 2, 1, 0, -1, -2, -3, -4, -5)$

Figure 5.12 Least-squares derivative approximation coefficients for $L = 2, 3, 4,$ and 5.

5.4.4 Second derivative

Figure 5.13 shows a simple filter for approximating the second derivative (Friesen et al., 1990), which has the difference equation

$$y(nT) = x(nT) - 2x(nT - 2T)) + x(nT - 4T) \tag{5.29}$$

This filter was derived by cascading two stages of the three-point central difference derivative of Eq. (5.26) and setting the amplitude multiplier to unity to obtain the transfer function

$$H(z) = (1 - z^{-2}) \times (1 - z^{-2}) = (1 - 2z^{-2} - z^{-4}) \tag{5.30}$$

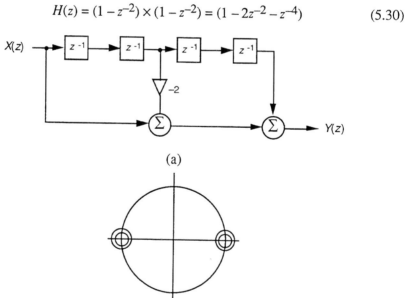

(a)

(b)

Figure 5.13 Second derivative. (a) Signal-flow graph. (b) Unit-circle diagram.

5.5 WINDOW DESIGN

A desired frequency response $H_d(\theta)$ (a continuous function) has as its inverse discrete-time Fourier transform (IDTFT), which is the desired unit pulse sequence, $h_d(k)$ (a discrete function). This sequence will have an infinite number of terms, so it is not physically realizable. The objective of window design is to choose an actual $h(k)$ with a finite number of terms such that the frequency response $H(e^{j\theta})$ will be in some sense close to $H_d(\theta)$. If the objective is to minimize the mean-squared error between the actual frequency response and the desired frequency response, then it can be shown by Parseval's theorem that the error is minimized by directly truncating $h_d(k)$. In other words, the pulse response $h(k)$ is chosen to be the first N terms of $h_d(k)$. Unfortunately, such a truncation results in large overshoots at sharp transitions in the frequency response, referred to as *Gibb's phenomenon*, illustrated in Figure 5.14.

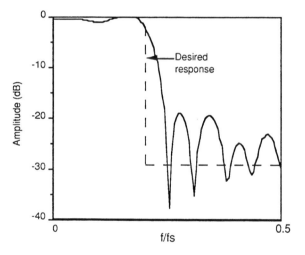

Figure 5.14 The overshoot that occurs at sharp transitions in the desired frequency response due to truncation of the pulse response is referred to as Gibb's phenomenon.

To understand this effect, consider that direct truncation of $h_d(k)$ is a multiplication of the desired unit pulse sequence by a rectangular window $w_R(k)$. Since multiplication in the time domain corresponds to convolution in the frequency domain, the frequency response of the resulting $h(k)$ is the convolution of the desired frequency response with the frequency response of the window function.

For example, consider that the frequency response of a rectangular window of infinite length is simply a unit pulse. The convolution of the desired frequency response with the unit pulse simply returns the desired response. However, as the width of the window decreases, its frequency response becomes less like an impulse function and sidelobes become more evident. The convolution of these sidelobes with the desired frequency response results in the overshoots at the transitions.

For a rectangular window, the response function $W_R(e^{j\theta})$ is given by

$$W_R(e^{j\theta}) = \frac{\sin\left(\frac{N\theta}{2}\right)}{\sin\left(\frac{\theta}{2}\right)}$$ (5.31)

where N is the length of the rectangular window. There are two important results of the convolution of the window with the desired frequency response. First the window function *smears* the desired response at transitions. This smearing of the transition bands increases the width of the transition from passband to stopband. This has the effect of increasing the width of the *main lobe* of the filter. Second, windowing causes undesirable ripple called *window leakage* or *sideband ripple* in the stopbands of the filter.

By using window functions other than a rectangular window, stopband ripple can be reduced at the expense of increasing main lobe width. Many types of windows have been studied in detail, including triangular (Bartlett), Hamming, Hanning, Blackman, Kaiser, Chebyshev, and Gaussian windows. Hanning and Hamming windows are of the form

$$w_H(k) = w_R(k)\left[\alpha + (1-\alpha)\cos\left(\frac{2\pi}{N}k\right)\right] \qquad \text{for } 0 < \alpha < 1 \quad (5.32)$$

where $\alpha = 0.54$ for the Hamming window and $\alpha = 0.50$ for the Hanning window.

The Kaiser window is of the form

$$w_K(k) = w_R(k)\, I_0\frac{\left(\alpha\sqrt{1-\left(\frac{k}{M}\right)^2}\right)}{I_0(\alpha)}$$ (5.33)

where α allows the designer to choose main lobe widths from the extreme minimum of the rectangular window to the width of the Blackman window, trading off sidelobe amplitude (Antoniou, 1979). See Roberts and Mullis (1987) for a discussion of the Chebyshev and Gaussian windows. All of these window functions taper to zero at the edges of the window. Figure 5.15 compares the performance of several of these windows.

Window Type	Sideband Ripple (dB)	Main Lobe Width
Rectangular	-13	$4\pi/N$
Triangular	-25	$8\pi/N$
Hanning	-31	$8\pi/N$
Hamming	-41	$8\pi/N$
Blackman	-38	$12\pi/N$

Figure 5.15 Responses of various windows.

Window design is usually performed iteratively, since it is difficult to predict the extent to which the transition band of the window will smear the frequency response. To design an FIR filter to meet a specific frequency response, one computes the unit pulse sequence of the desired response by taking the IDTFT, applies the window, and then computes the actual response by taking the DFT. The band edges or filter order are adjusted as necessary, and the process is repeated. Many algorithms have been developed to perform this process by computer. The technique can easily be constrained to generate filters with linear phase responses. Window designs do not generally yield the lowest possible order filter to meet the specifications. Windowing is also discussed in Chapter 11.

5.6 FREQUENCY SAMPLING

The frequency sampling method of design is more straightforward than the window design method since it circumvents the transformations from the time domain to the frequency domain. The terms in the filter response $H(e^{j\theta})$ are directly specified to match $H_d(\theta)$ at N uniformly spaced frequencies around the unit circle. As in the window design method, large overshoots will occur at sharp transitions in the response. This overshoot can be minimized by allowing some unconstrained terms in the transition band. Figure 5.16 illustrates a frequency sampling design with the same desired response and number of terms as in Figure 5.14, but with one unconstrained value chosen to minimize stopband ripple. The unconstrained values can be chosen to minimize some measure of error between $H(e^{j\theta})$ and $H_d(\theta)$. The frequency sampling method generally yields more efficient filters than the window method.

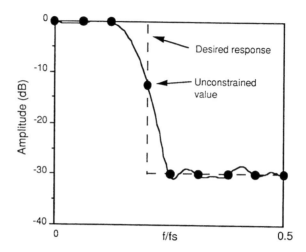

Figure 5.16 Frequency sampling design for $N = 16$ with one unconstrained term.

5.7 MINIMAX DESIGN

The window design method generates large errors at sharp transitions in the frequency response. Rather than minimizing the energy in the error

$$\varepsilon^2 = \frac{1}{2\pi} \int_{-\pi}^{\pi} \left| H_d(\theta) - H_1(e^{j\theta}) \right|^2 d\theta \qquad (5.34)$$

As in the window design method, it may be more desirable in certain applications to minimize the maximum error

$$\max_{\theta} \left| H_d(\theta) - H_1(e^{j\theta}) \right| \qquad (5.35)$$

The idea is to spread the error out evenly or in the more general case, in some weighted fashion across the frequency response. This is much more difficult than the window design problem, but algorithms have been developed to solve it by computer. The Remez algorithm, for example, allows the designer to select weighting factors throughout the passband and stopbands (Roberts and Mullis, 1987).

5.8 LAB: FIR FILTER DESIGN

The UW DigiScope software allows the user to experiment with several FIR design techniques, including direct specification of the pulse response, frequency sampling, window design, and placement of the zeros of the transfer equation. This lab studies some basic FIR filters such as the Hanning filter. It also uses specification of the pulse response and placement of the zeros of the transfer equation as design tools. Refer to Appendix D for information about using UW DigiScope.

5.8.1 Smoothing filters

Generate an ECG file with some random noise with the (G)`enwave` function, then run the Hanning filter by selecting (F)`ilters`, then (L)`oad filter`, then choosing `hanning.fil`. This function loads and immediately executes the filter function. By choosing (R)`un filter`, you can see the effect of cascading another Hanning filter with the first. The filter always operates on the data in the bottom channel.

The filters `poly2.fil`, `poly3.fil`, and `poly4.fil` are the least-squares smoothing filters with $L = 2$, 3, and 4 respectively. To test each of these three filters on the ECG data, first move the original signal from the top channel to the bottom channel with the (C)`opy data` command. Then use (L)`oad filter` to load and run the filters on ECG data. Observe the frequency and phase responses of each filter. How does changing the sampling rate of the ECG data affect the performance of the filters? Load `poly4.fil` and measure the time difference between the peak of the QRS complex in the unfiltered data and in the filtered data using the (M)`easure` command. You will need to change channels using (A)`ctive channel` to make the measurements. How do you account for the delay?

5.8.2 Derivatives

Experiment with the various derivative filters `deriv2.fil`, `deriv3.fil`, and `deriv5.fil` (which are the two-point, three-point central limit, and least-squares with $L = 2$ derivative filters respectively) on ECG and square wave data. Which filter is least suited for use with noisy signals?

5.8.3 Pulse Response

Create a filter that calculates the second derivative by selecting (F)`ilters`, then (D)`esign`, then (F)`IR`. Choose (P)`ulse resp`, and specify a filter length equal to 5. Enter the appropriate coefficients for the transfer equation and observe the frequency response. After saving the filter, return to the main menu, create a triangular wave with the (G)`enwave` function, and then use (R)`un filter` to observe the effects of the filter.

Create a linear phase filter with odd length (even order) and odd symmetry by satisfying Eq. (5.4). Observe its phase response.

5.8.4 Zero placement

The user may arbitrarily place zeros around the unit circle. The program automatically creates complex-conjugate pairs for values off the real axis. It then calculates the frequency, phase, and unit impulse responses for the filter.

Place zeros at $z = -0.5 \pm j0.866$ using the **(Z)ero place** function for a second-order filter, and comment on the frequency response. Run this filter on ECG data with and without 60-Hz noise sampled at 180 samples/s, and summarize the results.

5.8.5 Unit pulse sequence

There is a data file in the **STDLIB** called **ups.dat**. It consists of a single pulse. Reading this file and running an FIR filter will generate the unit impulse output sequence of the filter. Executing **(P)wr Spect** on this resulting output signal will give the magnitude response of the filter. In this way, the frequency response of filter can be measured more accurately than by the graph generated in the filter design utility. Use this method to find the 3-dB frequency of the Hanning filter (i.e., file **hanning.fil**).

5.8.6 Frequency sampling

Study the effect of increasing the width of the transition band by comparing the frequency response FIR filters of length 13 designed using frequency sampling by entering the following sets of values for the responses around the unit circle: {0, 0, –40, –40, –40, –40, –40} and {0, 0, –6, –40, –40, –40, –40}. Optimize the transition value (the third term in the sequence) for minimum stopband ripple.

Design a 60-Hz notch filter with length 21 by using frequency sampling. Run the filter on ECG data with and without 60-Hz noise. Does the filter perform better than **notch60.fil**? Optimize the filter for minimal gain at 60 Hz and minimal attenuation of other frequencies.

5.8.7 Window design

Compare the frequency response of low-pass filters designed with the various windows of the **(W)indow** function. Design the filters to have a cutoff frequency at 25 Hz with a sampling rate of 200 Hz. Make a table comparing main lobe width versus sideband ripple for the various windows. Measure the main lobe width at the first minimum in the frequency response.

5.8.8 Linear versus piecewise-linear phase response

Compare the phase response of the Hanning filter with that of the least-squares derivative filter with $L = 2$ (`deriv5.fil`). Why is the phase response of the least-squares filter not true linear? What causes the discontinuity? See Chapter 7 for additional information.

5.9 REFERENCES

Antoniou, A. 1979. *Digital Filters: Analysis and Design.* New York: McGraw-Hill.

Bogner, R. E. and Constantinides, A. G. 1985. *Introduction to Digital Filtering,* New York: John Wiley and Sons.

Friesen, G. M. et al. 1990. A comparison of the noise sensitivity of nine QRS detection algorithms, *IEEE Trans. Biomed. Eng.,* **BME-37**(1): 85–98.

Hamming, R. W. 1977. *Digital Filters.* Englewood Cliffs, NJ: Prentice Hall.

Rabiner, L. R. and Rader, C. M. 1972. *Digital Signal Processing* New York: IEEE Press.

Roberts, R. A. and Mullis, R. T. 1987. *Digital Signal Processing.* Reading, MA: Addison-Wesley.

5.10 STUDY QUESTIONS

5.1 What are the main differences between FIR and IIR filters?

5.2 What is the difference between direct convolution and "fast convolution"?

5.3 Why are finite-length register effects less significant in FIR filters than in IIR filters?

5.4 Compute and sketch the frequency response of a cascade of two Hanning filters. Does the cascade have linear phase?

5.5 Derive the phase response for an FIR filter with zeros located at $r\angle\pm\theta$ and $r^{-1}\angle\pm\theta$. Comment.

5.6 What are the trade-offs to consider when choosing the order of a least-squares polynomial smoothing filter?

5.7 Complete the derivation of coefficient values for the parabolic smoothing filter for $L = 2$.

5.8 Give some of the disadvantages of the simple 60-Hz notch filter described in section 5.3.1.

5.9 What are the main differences between the two-point difference and three-point central difference algorithms for approximating the derivative?

5.10 What are the three steps to designing a filter using the window method?

5.11 Explain the relationship between main-lobe width and sideband ripple for various windows.

5.12 How do the free parameters in the transition band of a frequency sampling design affect the performance of the filter?

5.13 What is the fundamental difference between minimax design and the window design method?

5.14 Design an FIR filter of length 15 with passband gain of 0 dB from 0 to 50 Hz and stopband attenuation of 40 dB from 100 to 200 Hz using the window design method. Compare the Hanning and rectangular windows. (Use a sampling rate of 400 sps.)

5.15 Repeat question 5.14 using frequency sampling.

5.16 The transfer function of the Hanning filter is

$$H_1(z) = \frac{1 + 2z^{-1} + z^{-2}}{4}$$

(a) What is its gain *at dc*? (b) Three successive stages of this filter are cascaded together to give a new transfer function [that is, $H(z) = H_1(z) \times H_1(z) \times H_1(z)$]. What is the overall gain of this filter *at dc*? (c) A high-pass filter is designed by subtracting the output of the Hanning filter from an all-pass filter with zero phase delay. How many zeros does the resulting filter have? Where are they located?

5.17 Two filters are cascaded. The first has the transfer function: $H_1(z) = 1 + 2z^{-1} - 3z^{-2}$. The second has the transfer function: $H_2(z) = 1 - 2z^{-1}$. A *unit impulse* is applied to the input of the cascaded filters. (a) What is the output sequence? (b) What is the magnitude of the amplitude response of the cascaded filter 1. at dc? 2. at 1/2 the foldover frequency? 3. at the foldover frequency?

5.18 Two filters are cascaded. The first has the transfer function: $H_1(z) = 1 + 2z^{-1} + z^{-2}$. The second has the transfer function: $H_2(z) = 1 - z^{-1}$. (a) A *unit impulse* is applied to the input of the cascaded filters. What is the output sequence? (b) What is the magnitude of the amplitude response of this cascaded filter at dc?

6

Infinite Impulse Response Filters

Ren Zhou

In this chapter we introduce the analysis and design of infinite impulse response (IIR) digital filters that have the potential of sharp rolloffs (Tompkins and Webster, 1981). We show a simple one-pole example illustrating the relationship between pole position and filter stability. Then we show how to design two-pole filters with low-pass, bandpass, high-pass, and band-reject characteristics. We also present algorithms to implement integrators that are all IIR filters. Finally, we provide a laboratory exercise that uses IIR filters for ECG analysis.

6.1 GENERIC EQUATIONS OF IIR FILTERS

The generic format of transfer function of IIR filters is expressed as the ratio of two polynomials:

$$H(z) = \frac{\sum\limits_{i=0}^{n} a_i z^{-i}}{1 - \sum\limits_{i=1}^{n} b_i z^{-i}} = \frac{a_0 + a_1 z^{-1} + a_2 z^{-2} + \ldots + a_n z^{-n}}{1 - b_1 z^{-1} - b_2 z^{-2} - \ldots - b_n z^{-n}} = \frac{Y(z)}{X(z)} \qquad (6.1)$$

Rearranging the terms gives

$$Y(z) = b_1 Y(z) z^{-1} + \ldots + b_n Y(z) z^{-n} + a_0 X(z) + a_1 X(z) z^{-1} + \ldots + a_n X(z) z^{-n} \quad (6.2)$$

The $Y(z)$ terms on the right side of this equation are delayed feedback terms. Figure 6.1 shows these feedback terms as recursive loops; hence, these types of filters are also called recursive filters.

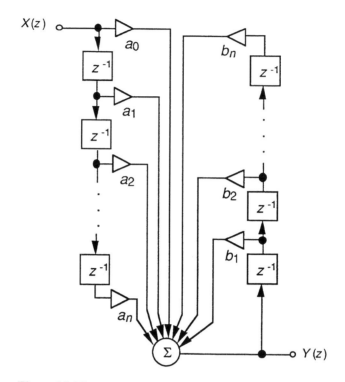

Figure 6.1 The output of an IIR filter is delayed and fed back.

6.2 SIMPLE ONE-POLE EXAMPLE

Let us consider the simple filter of Figure 6.2(a). We find the transfer function by applying a unit impulse sequence to the input $X(z)$. Figure 6.2(b) shows the sequences at various points in the filter for a feedback coefficient β equal to 1/2. Sequence (1) defines the unit impulse. From output sequence (2), we write the transfer function

$$H(z) = 1 + \frac{1}{2} z^{-1} + \frac{1}{4} z^{-2} + \frac{1}{8} z^{-3} + \ldots \qquad (6.3)$$

Using the binomial theorem, we write the infinite sum as a ratio of polynomials

$$H(z) = \frac{Y(z)}{X(z)} = \frac{1}{1 - \frac{1}{2} z^{-1}} \qquad (6.4)$$

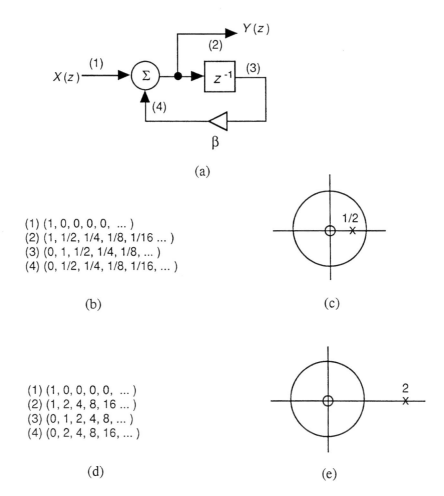

(1) (1, 0, 0, 0, 0, ...)
(2) (1, 1/2, 1/4, 1/8, 1/16 ...)
(3) (0, 1, 1/2, 1/4, 1/8, ...)
(4) (0, 1/2, 1/4, 1/8, 1/16, ...)

(b)

(c)

(1) (1, 0, 0, 0, 0, ...)
(2) (1, 2, 4, 8, 16 ...)
(3) (0, 1, 2, 4, 8, ...)
(4) (0, 2, 4, 8, 16, ...)

(d)

(e)

Figure 6.2 Simple one-pole recursive filters. (a) Block diagram. (b) Response to a unit pulse for $\beta = 1/2$. (c) Pole-zero plot for $\beta = 1/2$. (d) Response to a unit pulse for $\beta = 2$. (e) Pole-zero plot for $\beta = 2$.

We then rearrange this equation and write the output as a function of the feedback term and the input

$$Y(z) = \frac{1}{2} Y(z) \, z^{-1} + X(z) \tag{6.5}$$

Recognizing that $x(nT)$ and $y(nT)$ are points in the input and output sequences associated with the current sample time, they are analogous to the undelayed z-domain variables, $X(z)$ and $Y(z)$ respectively. Similarly $y(nT - T)$, the output value one sample point in the past, is analogous to the output z-domain output variable delayed by one sample point, or $Y(z)z^{-1}$. We can then write the difference equation by inspection

$$y(nT) = \frac{1}{2} y(nT - T) + x(nT) \qquad (6.6)$$

Unlike the FIR case where the current output $y(nT)$ is dependent only on current and past values of x, this IIR filter requires not only the current value of the input $x(nT)$ but also the previous value of the output itself $y(nT - T)$. Since past history of the output influences the next output value, which in turn influences the next successive output value, a transient requires a large number or sample points before it disappears from an output signal. As we have mentioned, this does not occur in an FIR filter because it has no feedback.

In order to find the poles and zeros, we first multiply the transfer function of Eq. (6.4) by z/z in order to make all the exponents of z positive.

$$H(z) = \frac{Y(z)}{X(z)} = \frac{1}{1 - \frac{1}{2} z^{-1}} \times \frac{z}{z} = \frac{z}{z - \frac{1}{2}} \qquad (6.7)$$

By equating the numerator to zero, we find that this filter has a single zero at $z = 0$. Setting the denominator equal to zero in order to locate the poles gives

$$z - \frac{1}{2} = 0$$

Thus, there is a single pole at $z = 1/2$.

The pole-zero plot for this single-pole filter is shown in Figure 6.2(c). In order to find the amplitude and phase responses, we substitute $z = e^{j\omega T}$ into the transfer function

$$H(\omega T) = \frac{1}{1 - \frac{1}{2} e^{-j\omega T}} = \frac{1}{\left[1 - \frac{1}{2} \cos(\omega T)\right] + j \left[\frac{1}{2} \sin(\omega T)\right]} \qquad (6.8)$$

Evaluating this function for the amplitude and phase responses, we find that this is a low-pass filter as illustrated in Figure 6.3.

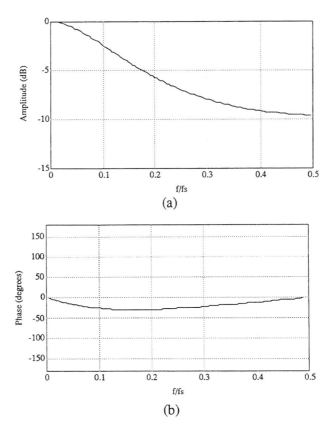

Figure 6.3 Simple one-pole recursive filter of Figure 6.2 with pole located at $z = 1/2$. (a) Amplitude response. (b) Phase response.

If we replace the multiplier constant β of 1/2 in this filter by 2, the output sequence in response to a unit impulse is (1, 2, 4, 8, 16, ...) as shown in Figure 6.2(d). The filter is unstable—the output increases with each successive sample. The response to a unit pulse input not only does not disappear with time, it increases by a factor of 2 each T s. We desire a unit pulse response that decays toward zero. Calculating the location of the pole with the multiplier of 2, we find that the pole is at $z = 2$, as shown in Figure 6.2(e). It is outside the unit circle, and the filter is unstable, as expected.

If we replace the multiplier constant of 1/2 in this filter by 1, the pole is directly on the unit circle at $z = 1$, and the output response to a unit impulse applied to the input is (1, 1, 1, 1, 1, ...). This filter is a special IIR filter, a rectangular integrator (see section 6.3).

6.3 INTEGRATORS

The general form of the integral is

$$A = \int_{t_1}^{t_2} f(t)dt \tag{6.9}$$

where A is the area under the function between the limits t_1 to t_2. The digital implementation for the determination of integral solutions is done by approximating the function using curve fitting techniques at a finite number of points, where

$$A = \sum_{n=t_1}^{t_2} f(n)\Delta t \tag{6.10}$$

The Laplace transform of an integrator is

$$H(s) = \frac{1}{s} \tag{6.11}$$

The amplitude and phase responses are found by substituting into the transfer function the relation, $s = j\omega$, giving

$$H(j\omega) = \frac{1}{j\omega} \tag{6.12}$$

Thus, the ideal amplitude response is inversely proportional to frequency

$$|H(j\omega)| = \left|\frac{1}{\omega}\right| \tag{6.13}$$

and the phase response is

$$\angle H(j\omega) = \tan^{-1}\left(\frac{-1/\omega}{0}\right) = -\frac{\pi}{2} \tag{6.14}$$

A number of numerical techniques exist for digital integration. Unlike the analog integrator, a digital integrator has no drift problems because the process is a computer program, which is not influenced in performance by residual charge on capacitors. We discuss here three popular digital integration techniques—rectangular summation, trapezoidal summation, and Simpson's rule.

6.3.1 Rectangular integration

This algorithm performs the simplest integration. It approximates the integral as a sum of rectangular areas. Figure 6.4(a) shows that each rectangle has a base equal in length to one sample period T and in height to the value of the most recently sampled input $x(nT - T)$. The area of each rectangle is $Tx(nT - T)$. The difference equation is

$$y(nT) = y(nT - T) + T\,x(nT) \qquad (6.15)$$

where $y(nT - T)$ represents the sum of all the rectangular areas prior to adding the most recent one. The error in this approximation is the difference between the area of the rectangle and actual signal represented by the sampled data. The z transform of Eq. (6.15) is

$$Y(z) = Y(z)\,z^{-1} + T\,X(z) \qquad (6.16)$$

and

$$H(z) = \frac{Y(z)}{X(z)} = T\left(\frac{1}{1 - z^{-1}}\right) \qquad (6.17)$$

Figure 6.4(a) shows that this transfer function has a pole at $z = 1$ and a zero at $z = 0$. The amplitude and phase responses are

$$|H(\omega T)| = \left|\frac{T}{2\sin(\omega T/2)}\right| \qquad (6.18)$$

and

$$\angle H(\omega T) = \frac{\omega T}{2} - \frac{\pi}{2} \qquad (6.19)$$

Figure 6.5 shows the amplitude response. We can reduce the error of this filter by increasing the sampling rate significantly higher than the highest frequency present in the signal that we are integrating. This has the effect of making the width of each rectangle small compared to the rate of change of the signal, so that the area of each rectangle better approximates the input data. Increasing the sampling rate also corresponds to using only the portion of amplitude response at the lower frequencies, where the rectangular response better approximates the ideal response. Unfortunately, higher than necessary sampling rates increase computation time and waste memory space by giving us more sampled data than are necessary to characterize a signal. Therefore, we select other digital integrators for problems where higher performance is desirable.

132   Biomedical Digital Signal Processing

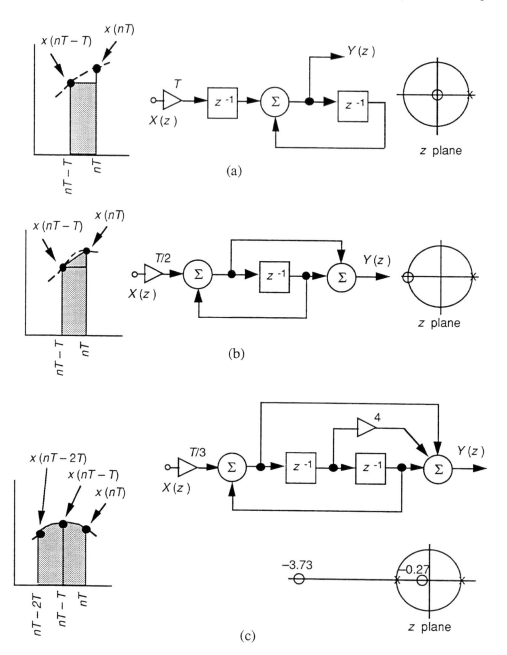

Figure 6.4 Integration. (a) Rectangular. (b) Trapezoidal. (c) Simpson's rule.

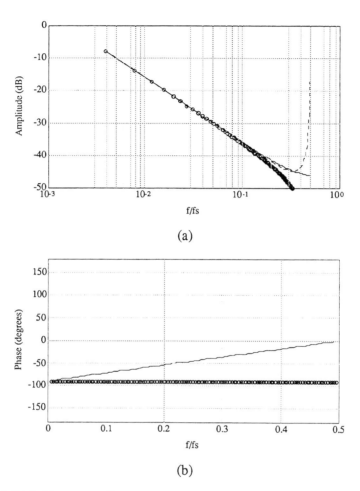

(a)

(b)

Figure 6.5 Digital integrators. The sampling period T is set equal to 1. (a) Amplitude responses. (b) Phase responses. Solid line: rectangular. Circles: trapezoidal. Dashed line: Simpson's rule.

6.3.2 Trapezoidal integration

This filter improves on rectangular integration by adding a triangular element to the rectangular approximation, as shown in Figure 6.4(b). The difference equation is the same as for rectangular integration except that we add the triangular element.

$$y(nT) = y(nT - T) + Tx(nT - T) + \frac{T}{2}[x(nT) - x(nT - T)]$$

$$= y(nT - T) + \frac{T}{2}[x(nT) + x(nT - T)] \qquad (6.20)$$

This corresponds to the transfer function

$$H(z) = \frac{T}{2}\left(\frac{1 + z^{-1}}{1 - z^{-1}}\right) \qquad (6.21)$$

For the exact amplitude and phase responses, we use the analytical approach. Since we desire the frequency response, we evaluate the transfer function on the unit circle by simply substituting $e^{j\omega T}$ for every occurrence of z. For the recursive filter

$$H(z) = \frac{1 + z^{-1}}{1 - z^{-1}} \qquad (6.22)$$

we obtain

$$H(\omega T) = \frac{1 + e^{-j\omega T}}{1 - e^{-j\omega T}} = \frac{e^{-j\frac{\omega T}{2}}\left(e^{j\frac{\omega T}{2}} + e^{-j\frac{\omega T}{2}}\right)}{e^{-j\frac{\omega T}{2}}\left(e^{j\frac{\omega T}{2}} - e^{-j\frac{\omega T}{2}}\right)} = \frac{2\cos\left(\frac{\omega T}{2}\right)}{2j\sin\left(\frac{\omega T}{2}\right)} = -j\cot\left(\frac{\omega T}{2}\right)$$

The amplitude response is the magnitude of $H(\omega T)$

$$|H(\omega T)| = \left|\cot\left(\frac{\omega T}{2}\right)\right| \qquad (6.23)$$

and the phase response is

$$\angle H(\omega T) = -\frac{\pi}{2} \qquad (6.24)$$

The block diagram and pole-zero plot are shown in Figure 6.4(b). Like the rectangular approach, this function still has a pole at $z = 1$, but the zero is moved to $z = -1$, the location of the folding frequency, giving us zero amplitude response at f_0. The amplitude and phase response are

$$|H(\omega T)| = \left|\frac{T}{2}\cot\left(\frac{\omega T}{2}\right)\right| \qquad (6.25)$$

and

$$\angle H(\omega T) = -\frac{\pi}{2} \tag{6.26}$$

The amplitude response approximates that of the ideal integrator much better than the rectangular filter and the phase response is exactly equal to that of the ideal response. The trapezoidal technique provides a very simple, effective recursive integrator.

Figure 6.6 shows a program that performs trapezoidal integration by direct implementation of the difference equation, Eq. (6.20).

```
/*******************************************************************
 *   Turbo C source code implementing a trapezoidal integrator
 *
 *   Assume sampling period is 2 ms, the difference equation is:
 *
 *        y(nT) = y(nT - T) + 0.001x(nT) + 0.001x(nT - T)
 *
 *******************************************************************/
float Trapezoid(signal)
float signal;                    /* x[nT] */
{
  static float  x[2], y[2];
  int count;

  x[1]=x[0];                     /* x[1]=x(nT-T)  */
  x[0]=signal;                   /* x[0]=x(nT)  */

  y[0]=y[1]+0.001*x[0]+0.001*x[1];/* difference equation */

  y[1]=y[0];                     /* y[1]=y(nT - T)  */

  return y[0];
}
```

Figure 6.6 A trapezoidal integrator written in the C language.

6.3.3 Simpson's rule integration

Simpson's rule is the most widely used numerical integration algorithm. Figure 6.4(c) shows that this approach approximates the signal corresponding to three input sequence points by a polynomial fit. The incremental area added for each new input point is the area under this polynomial. The difference equation is

$$y(nT) = y(nT - 2T) + \frac{T}{3}[x(nT) + 4x(nT - T) + x(nT - 2T)] \tag{6.27}$$

The transfer function is

$$H(z) = \frac{T}{3}\left(\frac{1 + 4z^{-1} + z^{-2}}{1 - z^{-2}}\right)$$ (6.28)

Figure 6.4(c) shows the block diagram and pole-zero plot for this filter. The amplitude and phase responses are

$$|H(\omega T)| = \left|\frac{T}{2}\left(\frac{2 + \cos \omega T}{\sin \omega T}\right)\right|$$ (6.29)

and

$$\angle H(\omega T) = -\frac{\pi}{2}$$ (6.30)

Figure 6.5 shows the amplitude response of the Simpson's rule integrator. Like the trapezoidal technique, the phase response is the ideal $-\pi/2$. Simpson's rule approximates the ideal integrator for frequencies less than about $f_s/4$ better than the other techniques. However, it amplifies high-frequency noise near the foldover frequency. Therefore, it is a good approximation to the integral but is dangerous to use in the presence of noise. Integration of noisy signals can be accomplished better with trapezoidal integration.

6.4 DESIGN METHODS FOR TWO-POLE FILTERS

IIR filters have the potential for sharp rolloffs. Unlike the design of an FIR filter, which frequently is based on approximating an input sequence numerically, IIR filter design frequently starts with an analog filter that we would like to approximate. Their transfer functions can be represented by an infinite sum of terms or ratio of polynomials. Since they use feedback, they may be unstable if improperly designed. Also, they typically do not have linear phase response.

6.4.1 Selection method for r and θ

The general design equation has a standard recursive form for the four types of filters: low-pass, bandpass, high-pass, and band-reject:

$$H(z) = \frac{1 + a_1 z^{-1} + a_2 z^{-2}}{1 - b_1 z^{-1} + b_2 z^{-2}}$$ (6.31)

where the zero locations are

$$z = \frac{-a_1 \pm \sqrt{a_1{}^2 - 4a_2}}{2} \qquad (6.32)$$

and the pole locations are

$$z = \frac{b_1 \pm \sqrt{b_1{}^2 - 4b_2}}{2} \qquad (6.33)$$

where

$$b_1 = 2r\cos\theta \qquad \text{and} \qquad b_2 = r^2 \qquad (6.34)$$

Also

$$\theta = 2\pi \left(\frac{f_c}{f_s}\right) \qquad (6.35)$$

Figure 6.7 shows the structure of the two-pole filter. Assigning values to the numerator coefficients, a_1 and a_2, as shown in Figure 6.8 establishes the placement of the two zeros of the filter and defines the filter type. Values of b_1 and b_2 are determined by placing the poles at specific locations within the unit circle.

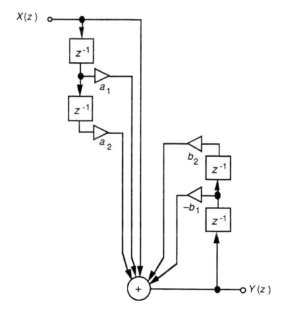

Figure 6.7 Block diagram showing two-pole filter structure.

We start the design by selecting the sampling frequency f_s, which must be at least twice the highest frequency contained in the input signal. Next we choose the critical frequency f_c. This is the cutoff frequency for low- and high-pass filters, the resonant frequency for a bandpass filter, and the notch frequency for a band-reject filter. These two choices establish θ, the angular location of the poles.

We then select r, the distance of the poles from the origin. This is the damping factor, given by

$$r = e^{-aT} \tag{6.36}$$

We know from the amplitude response that underdamping occurs as the poles approach the unit circle (i.e., $r \rightarrow 1$ or $a \rightarrow 0$) and overdamping results for poles near the origin (i.e., $r \rightarrow 0$ or $a \rightarrow \infty$). Moving the poles by increasing r or θ causes underdamping, and decreasing either variable causes overdamping (Soderstrand, 1972). As we have done in other designs, we find the frequency response by substituting $e^{j\omega T}$ for z in the final transfer function.

	a_1	a_2
Low-pass	2	1
Bandpass	0	−1
High-pass	−2	1
Band-reject	$2\cos\theta$	1

Figure 6.8 Table of numerator coefficients for the two-pole filter implementations of Eq. (6.31).

We can write the difference equation for the two-pole filter by first rewriting Eq. (6.31):

$$H(z) = \frac{Y(z)}{X(z)} = \frac{1 + a_1 z^{-1} + a_2 z^{-2}}{1 - b_1 z^{-1} + b_2 z^{-2}} \tag{6.37}$$

Then rearranging terms to find $Y(z)$, we get

$$Y(z) = b_1 Y(z)z^{-1} - b_2 Y(z)z^{-2} + X(z) + a_1 X(z)z^{-1} + a_2 X(z)z^{-2} \tag{6.38}$$

We can now directly write the difference equation using analogous discrete variable terms

$$y(nT) = b_1\, y(nT - T) - b_2\, y(nT - 2T) + x(nT)$$
$$+ a_1\, x(nT - T) + a_2\, x(nT - 2T) \tag{6.39}$$

Figure 6.9 shows the pole-zero plots and corresponding rubber membrane representations for each of the four filter types.

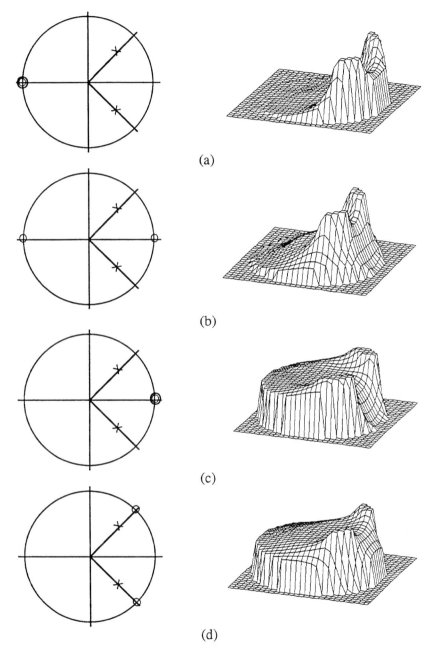

(a)

(b)

(c)

(d)

Figure 6.9 Pole-zero plots and rubber membrane views of elementary two-pole recursive digital filters. (a) Low-pass. (b) Bandpass. (c) High-pass. (d) Band-reject (notch).

Let us now design a filter and write a C-language program to implement it. First we choose a low-pass filter. This establishes the locations of the two zeros and establish the values for $a_1 = 2$ and $a_2 = 1$ (see Figure 6.8). Then we decide on the locations of the poles. For this design, we choose locations at $r = 0.5$ and $\theta = \pm45°$. This establishes the values for $b_1 = 0.707$ and $b_2 = 0.25$ [see Eq. (6.34)].

Software implementation of this digital filter is now fairly straightforward. We write a filter subroutine directly from the difference equation. Figure 6.10 shows how this low-pass filter is executed in five steps:

1. Pass in the current input signal.
2. Shift array elements to get delays of past input signal and store them.
3. Accumulate all past and/or current input and past output terms with multiplying coefficients.
4. Shift array elements to get delays of past output signal and store them.
5. Return with output signal.

```
/************************************************************
 *    Turbo C source code for implementing an IIR low-pass filter
 *
 *    Assume we use the r and q method where r = 0.5, q = 45°.
 *    The difference equation is:
 *
 *    y(nT) = 0.707y(nT-T)-0.25y(nT-2T)+x(nT)+2x(nT-T)+x(nT-2T)
 *    y(nT) = 0.707*y[1] - 0.25*y[2] + x[0] + 2*x[1] + x[2]
 *
 ************************************************************/
float LowPass(signal)
float signal;                    /* x[nT] */
{
  static float  x[3], y[3];
  int count;

  for ( count=2; count>0; count-- )
    x[count]=x[count-1];                    /* shift for x term delays */

  x[0]=signal;                        /* x[0] = x[nT] */
                                      /* difference equation */

  y[0]=0.707*y[1]-0.25*y[2]+x[0]+2.0*x[1]+x[2];

  for( count = 2; count > 0; count--)
    y[count]=y[count-1];            /* shift for y term delays */

  return y[0];
}
```

Figure 6.10 An IIR low-pass filter written in the C language.

The same algorithm can be used for high-pass, bandpass, and band-reject filters as well as for integrators. The only disadvantage of the software implementation is its limited speed. However, this program will execute in real time on a modern PC with a math coprocessor for biomedical signals like the ECG.

These two-pole filters have operational characteristics similar to second-order analog filters. The rolloff is slow, but we can cascade several identical sections to improve it. Of course, we may have time constraints in a real-time system which limit the number of sections that can be cascaded in software implementations. In these cases, we design higher-order filters.

6.4.2 Bilinear transformation method

We can design a recursive filter that functions approximately the same way as a model analog filter. We start with the s-plane transfer function of the filter that we desire to mimic. We accomplish the basic digital design by replacing all the occurrences of the variable s by the approximation

$$s \approx \frac{2}{T} \left[\frac{1 - z^{-1}}{1 + z^{-1}} \right] \tag{6.40}$$

This substitution does a nonlinear translation of the points in the s plane to the z plane. This warping of the frequency axis is defined by

$$\omega' = \frac{2}{T} \left[\tan \left(\frac{\omega T}{2} \right) \right] \tag{6.41}$$

where ω' is the analog domain frequency corresponding to the digital domain frequency ω. To do a bilinear transform, we first prewarp the frequency axis by substituting the relation for ω' for all the critical frequencies in the Laplace transform of the filter. We then replace s in the transform by its z-plane equivalent. Suppose that we have the transfer function

$$H(s) = \frac{\omega_c'}{s^2 + \omega_c'^2} \tag{6.42}$$

We substitute for ω_c' using Eq. (6.41) and for s with Eq. (6.40), to obtain

$$H(z) = \frac{\dfrac{2}{T} \left[\tan \left(\dfrac{\omega T}{2} \right) \right]}{\left\{ \dfrac{2}{T} \left[\dfrac{1 - z^{-1}}{1 + z^{-1}} \right] \right\}^2 + \left\{ \dfrac{2}{T} \left[\tan \left(\dfrac{\omega T}{2} \right) \right] \right\}^2} \tag{6.43}$$

This z transform is the description of a digital filter that performs approximately the same as the model analog filter. We can design higher-order filters with this technique.

6.4.3 Transform tables method

We can design digital filters to approximate analog filters of any order with filter tables such as those in Stearns (1975). These tables give the Laplace and z-transform equivalents for corresponding continuous and discrete-time functions. To illustrate the design procedure, let us consider the second-order filter of Figure 6.11(a). The analog transfer function of this filter is

$$H(s) = \frac{A}{LC}\left[\frac{1}{s^2 + (R/L)s + 1/LC}\right] \qquad (6.44)$$

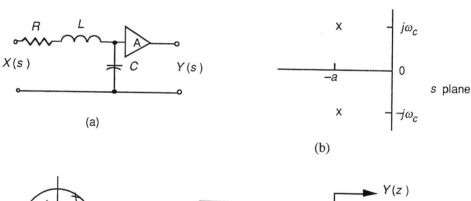

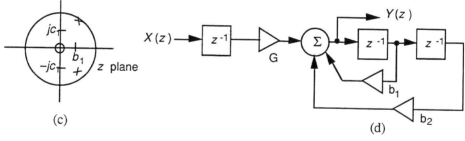

Figure 6.11 Second-order filter. (a) Analog filter circuit. (b) Transfer function pole-zero plot for the analog filter. (c) Pole-zero plot for digital version of the second-order filter. (d) Block diagram of the digital filter.

Solving for the poles, we obtain

$$s = -a \pm j\omega_c \tag{6.45}$$

where

$$a = \frac{R}{2L} \tag{6.46}$$

$$\omega_c = \left[\frac{1}{LC} - \frac{R^2}{4L^2}\right]^{1/2} \tag{6.47}$$

We can rewrite the transfer function as

$$H(s) = \frac{A}{LC}\left[\frac{1}{(s+a)^2 + \omega_c^2}\right] \tag{6.48}$$

Figure 6.11(b) shows the s-plane pole-zero plot. This s transform represents the continuous time function $e^{-at} \sin\omega_c t$. The z transform for the corresponding discrete-time function $e^{-naT} \sin n\omega_c T$ is in the form

$$H(z) = \frac{Gz^{-1}}{1 - b_1 z^{-1} - b_2 z^{-2}} \tag{6.49}$$

where

$$b_1 = 2e^{-aT} \cos\omega_c T \tag{6.50}$$

and

$$b_2 = -e^{-2aT} \tag{6.51}$$

Also

$$G = \frac{A}{\omega_c LC} e^{-aT} \sin\omega_c T \tag{6.52}$$

Variables a, ω_c, A, L, and C come from the analog filter design. This transfer function has one zero at $z = 0$ and two poles at

$$z = b_1 \pm j(b_1^2 + 4b_2)^{1/2} = b_1 \pm jc_1 \tag{6.53}$$

Figure 6.11(c) shows the z-plane pole-zero plot. We can find the block diagram by substituting the ratio $Y(z)/X(z)$ for $H(z)$ and collecting terms.

$$Y(z) = GX(z)z^{-1} + b_1 Y(z)z^{-1} + b_2 Y(z)z^{-2} \tag{6.54}$$

The difference equation is

$$y(nT) = Gx(nT - T) + b_1 y(nT - T) + b_2 y(nT - 2T) \tag{6.55}$$

From this difference equation we can directly write a program to implement the filter. We can also construct the block diagram as shown in Figure 6.11(d).

This transform-table design procedure provides a technique for quickly designing digital filters that are close approximations to analog filter models. If we have the transfer function of an analog filter, we still usually must make a substantial effort to implement the filter with operational amplifiers and other components. However, once we have the z transform, we have the complete filter design specified and only need to write a straightforward program to implement the filter.

6.5 LAB: IIR DIGITAL FILTERS FOR ECG ANALYSIS

This lab provides experience with the use of IIR filters for processing ECGs. In order to design integrators and two-pole filters using UW DigiScope, select (F)ilters from the main menu, choose (D)esign, then (I)IR.

6.5.1 Integrators

This chapter reviewed three different integrators (rectangular, trapezoidal, and Simpson's rule). Which of these filters requires the most computation time? What do the unit-circle diagrams of these three filters have in common?

Run the rectangular integrator on an ECG signal (e.g., ecg105.dat). Explain the result. Design a preprocessing filter to solve any problem observed. Compare the output of the three integrators on appropriately processed ECG data with and without random noise. Which integrator is best to use for noisy signals?

6.5.2 Second-order recursive filters

Design three two-pole bandpass filters using $r = 0.7, 0.9,$ and 0.95 for a critical frequency of 17 Hz (sampling rate of 200 sps). Measure the Q for the three filters. Q is defined as the ratio of critical frequency to the 3-dB bandwidth (difference between the 3-dB frequencies). Run the three filters on an ECG file and contrast the outputs. What trade-offs must be considered when selecting the value of r?

6.5.3 Transfer function (Generic)

The (G)eneric tool allows you to enter coefficients of a transfer function of the form

$$H(z) = \frac{a_0 + a_1 z^{-1} + a_2 z^{-2} + \ldots + a_n z^{-n}}{1 + b_1 z^{-1} + b_2 z^{-2} + \ldots + b_m z^{-m}} \qquad \text{where } m <= n$$

Using the example from the transform tables method discussed in section 6.4.3 with $R = 10\ \Omega$, $L = 10$ mH, and $C = 1\ \mu$F, calculate and enter the coefficients of the resulting $H(z)$. Use a sampling rate of 200 sps. What is the Q of this filter?

6.5.4 Pole-zero placement

Create a 60-Hz notch FIR filter for a sampling rate of 180 sps by placing two zeros on the unit circle (i.e., $r = 1.0$) at angles of $\pm120°$ as shown in Figure 6.12(a). Create an IIR filter by adding poles inside the circle at angles of $\pm110°$ and $\pm130°$ at $r = 0.9$, as shown in Figure 6.12(b). Compare the amplitude response of this filter with that of the FIR filter. Measure the Q of both filters. Can you further increase the Q of the IIR filter? What happens to the phase response?

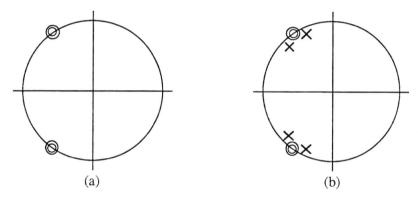

(a)   (b)

Figure 6.12 Notch filters. (a) FIR filter. (b) IIR filter.

6.6 REFERENCES

Soderstrand, M. A. 1972. On-line digital filtering using the PDP-8 or PDP-12. *Computers in the Neurophysiology Laboratory*, **1**: 31-49. Maynard, MA: Digital Equipment Corporation.
Stearns, S. D. 1975. *Digital Signal Analysis*. Rochelle Park, NJ: Hayden.
Tompkins, W. J. and Webster, J. G. (eds.) 1981. *Design of Microcomputer-based Medical Instrumentation*. Englewood Cliffs, NJ: Prentice Hall.

6.7 STUDY QUESTIONS

6.1 Describe the characteristics of the generic transfer function of recursive filters.
6.2 What is the difference between IIR filters and recursive filters?
6.3 How do you derive the frequency response of a recursive filter?

6.4 How do you write the difference equation from the transfer function of a recursive filter?
6.5 Design a low-pass recursive filter using the r and θ method.
6.6 Design a high-pass filter using bilinear transformation method.
6.7 Design a filter using transform tables method.
6.8 A digital filter has a *unit impulse* output sequence {i.e., 1, 0, 0,...} when a *unit step* {i.e., 1, 1, 1, 1,...} is applied at its input. What is the transfer function of the filter?
6.9 Write a rectangular integrator in C.
6.10 Using the difference equation from the result of question 6.5, run the filtering program provided in section 6.4.
6.11 Draw the z-plane pole-zero plot for a filter with the z transform:

$$H(z) = \frac{1 + 2z^{-1} + z^{-2}}{1 - z^{-2}}$$

6.12 A filter has the following output sequence in response to a unit impulse: {−2, 4, −8, 16, ...}. Write its z transform in closed form (i.e., as a ratio of polynomials). From the following list, indicate all the terms that describe this filter: recursive, nonrecursive, stable, unstable, FIR, IIR.
6.13 A digital filter has the following transfer function:

$$H(z) = \frac{1 - bz^{-1}}{1 + bz^{-1}}$$

(a) What traditional type of filter is this if b = (1) 0.8; (2) −0.8; (3) 1; (4) −2; (5) −1/2?
(b) If $b = -1/2$, what is the filter gain? (c) If $b = 1/2$, what is the difference equation for $y(nT)$?
6.14 The block diagrams for four digital filters are shown below. Write their (a) transfer functions, (b) difference equations.

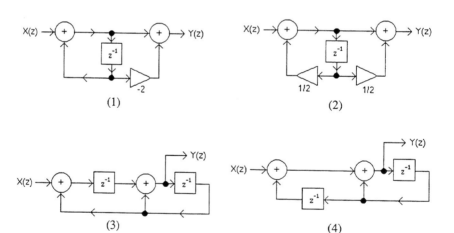

(1)   (2)

(3)   (4)

6.15 Write the transfer function and comment on the stability of a filter with the following output sequence in response to a unit impulse: (a) {3, −6, 12, −24, ...}. (b) {2, −1, 1, −1/2, 1/4, −1/8, ...}.
6.16 A digital filter has a difference equation: $y(nT) = y(nT - 2T) + x(nT - T)$. What is its output sequence in response to a unit impulse applied to its input?

6.17 A filter's difference equation is: $y(nT) = x(nT) + 3y(nT - T)$. What is its output sequence in response to a unit impulse?

6.18 The difference equation of a filter is: $y(nT) = x(nT) + x(nT - 2T) + y(nT - 2T)$. Where are its poles and zeros located?

6.19 The general equation for a two-pole digital filter is:

$$H(z) = \frac{1 + a_1 z^{-1} + a_2 z^{-2}}{1 - b_1 z^{-1} + b_2 z^{-2}}$$

where $b_1 = 2r\cos(\theta)$ and $b_2 = r^2$

(a) What traditional type of filter is this if (1) $a_1 = 2$, $a_2 = 1$, $-2 < b_1 < 2$, $0 < b_2 < 1$, (2) $a_1 = -2\cos(\theta)$, $a_2 = 1$, $r = 1$, $\theta = 60°$? (b) What is the difference equation for $y(nT)$ if $a_1 = 0$, $a_2 = -1$, $r = 1/2$, $\theta = 60°$?

6.20 A digital filter has two zeros located at $z = 0.5$ and $z = 1$, and two poles located at $z = 0.5$ and $z = 0$. Write an expression for (a) its amplitude response as a function of a single trigonometric term, and (b) its phase response.

6.21 The block diagram of a digital filter is shown below.
(a) Consider the case when the switch is **open**. (1) What is the filter's transfer function? (2) Write its difference equation. (3) What is the magnitude of its amplitude response at one-half the sampling frequency?
(b) Consider the case when the switch is **closed**. (1) What is the filter's transfer function? (2) Write its difference equation. (3) If G is a negative fraction, what traditional filter type best describes this filter? (4) If $G = 1$, where are its poles and zeros located? (5) If a unit-amplitude step function is applied to the input with $G = 1$, what is the output sequence? (6) If a unit-amplitude step function is applied to the input with $G = -1$, what is the output sequence?

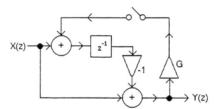

6.22 A digital filter has the block diagram shown below. (a) Write its transfer function. (b) Where are its poles and zeros located?

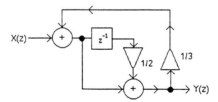

6.23 The difference equation for a filter is: $y(nT) = 2y(nT - T) + 2x(nT) + x(nT - T)$. Draw its z-plane pole-zero plot, indicating the locations of the poles and zeros.

6.24 Application of a unit **impulse** to the input of a filter produces the output sequence {1, 0, 1, 0, 1, 0, ...}. What is the difference equation for this filter?

6.25 Application of a unit **step** to the input of a filter produces the output sequence {1, 1, 2, 2, 3, 3, ...}. What is the difference equation for this filter? HINT: The z transform of a unit step is

$$\frac{1}{1-z^{-1}}$$

6.26 Write the transfer functions of the following digital filters:

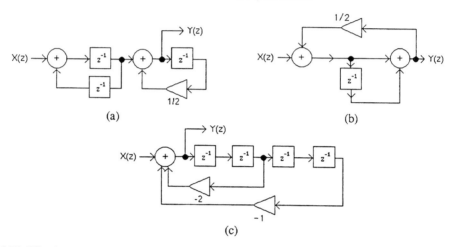

(a)

(b)

(c)

6.27 What is the phase response for a digital filter with the transfer function

$$H(z) = \frac{1-z^{-6}}{1+z^{-6}}$$

6.28 Write the amplitude response of a filter with the transfer function:

$$H(z) = \frac{z^{-2}}{1-z^{-2}}$$

6.29 A filter has two poles at $z = 0.5 \pm j0.5$ and two zeros at $z = 0.707 \pm j0.707$. What traditional filter type best describes this filter?

6.30 A filter operating at a sampling frequency of 1000 samples/s has a pole at $z = 1/2$ and a zero at $z = 3$. What is the magnitude of its amplitude response at **dc**?

6.31 A filter is described by the difference equation: $y(nT) = x(nT) + x(nT - T) - 0.9y(nT - T)$. What is its transfer function?

6.32 A filter has the difference equation: $y(nT) = y(nT - 2T) + x(nT) + x(nT - T)$. What traditional filter type **best** describes this filter?

6.33 In response to a unit impulse applied to its input, a filter has the output sequence: {1, 1/2, 1/4, 1/8,...}. What is its transfer function?

6.34 The difference equation for a digital filter is: $y(nT) = x(nT) - ax(nT - T) - by(nT - T)$. Variables a and b are positive integers. What traditional type of filter is this if $a = 1$ and (a) $b = 0.8$, (b) $b > 1$?

6.35 Write the (a) amplitude response, (b) phase response, and (c) difference equation for a filter with the transfer function:

$$H(z) = \frac{1 - z^{-1}}{1 + z^{-1}}$$

6.36 Write the (a) amplitude response, (b) phase response, and (c) difference equation for a filter with the transfer function:

$$H(z) = \frac{z - 1}{2z + 1}$$

6.37 A filter operating at a sampling frequency of 1000 samples/s has a pole at $z = 1$ and a zero at $z = 2$. What is the magnitude of its amplitude response at 500 Hz?

6.38 A filter operating at a sampling frequency of 200 samples/s has poles at $z = \pm j/2$ and zeros at $z = \pm 1$. What is the magnitude of its amplitude response at 50 Hz?

6.39 A filter is described by the difference equation: $2y(nT) + y(nT - T) = 2x(nT)$. What is its transfer function $H(z)$?

6.40 A filter has the difference equation: $y(nT) = y(nT - T) - y(nT - 2T) + x(nT) + x(nT - T)$. What is its transfer function?

6.41 In response to a unit impulse applied to its input, a filter has the output sequence: $\{1, 1, 1/2, 1/4, 1/8,...\}$. What is its transfer function?

6.42 A filter has a transfer function that is identical to the z transform of a unit step. A unit step is applied at its input. What is its output sequence?

6.43 A filter has a transfer function that is equal to the z transform of a ramp. A unit impulse is applied at its input. What is its output sequence? HINT: The equation for a ramp is $x(nT) = nT$, and its z transform is

$$X(z) = \frac{Tz^{-1}}{(1 - z^{-1})^2}$$

6.44 A ramp applied to the input of digital filter produces the output sequence: $\{0, T, T, T, T, ...\}$. What is the transfer function of the filter?

6.45 A digital filter has a unit step (i.e., 1, 1, 1, 1, ...) output sequence when a unit impulse (i.e., 1, 0, 0, ...) is applied at its input. How is this filter **best** described?

6.46 A discrete impulse function is applied to the inputs of four different filters. For each of the output sequences that follow, state whether the filter is recursive or nonrecursive. (a) $\{1, 2, 3, 4, 5, 6, 0, 0, 0,...\}$, (b) $\{1, -1, 1, -1, 1, -1,...\}$, (c) $\{1, 2, 4, 8, 16,...\}$, (d) $\{1, 0.5, 0.25, 0.125,...\}$.

6.47 What similarities are common to all three integrator algorithms discussed in the text (i.e., rectangular, trapezoidal, and Simpson's rule)?

6.48 A differentiator is cascaded with an integrator. The differentiator uses the two-point difference algorithm:

$$H_1(z) = \frac{1 - z^{-1}}{T}$$

The integrator uses trapezoidal integration:

$$H_2(z) = \frac{T}{2} \left[\frac{1 + z^{-1}}{1 - z^{-1}} \right]$$

A unit impulse is applied to the input. What is the output sequence?

6.49 A differentiator is cascaded with an integrator. The differentiator uses the three-point central difference algorithm:

$$H_1(z) = \frac{1 - z^{-2}}{2T}$$

The integrator uses rectangular integration:

$$H_2(z) = T\left[\frac{1}{1 - z^{-1}}\right]$$

(a) A unit impulse is applied to the input. What is the output sequence? (b) What traditional filter type best describes this filter?

6.50 A digital filter has two zeros located at $z = 0.5$ and $z = 1$, and a single pole located at $z = 0.5$. Write an expression for (a) its amplitude response as a function of a single trigonometric term, and (b) its phase response.

6.51 In response to a unit impulse applied to its input, a filter has the output sequence: $\{2, -1, 1, -1/2, 1/4, -1/8, \ldots\}$. What is its transfer function?

6.52 The difference equation for a filter is: $y(nT - T) = x(nT - T) + 2x(nT - 4T) + 4x(nT - 10T)$. What is its transfer function, $H(z)$?

6.53 What is the transfer function $H(z)$ of a digital filter with the difference equation

$$y(nT - 2T) = y(nT - T) + x(nT - T) + x(nT - 4T) + x(nT - 10T)$$

6.54 A digital filter has the following output sequence in response to a *unit impulse*: $\{1, -2, 4, -8, \ldots\}$. Where are its poles located?

6.55 A digital filter has a single zero located at $z = 0.5$ and a single pole located at $z = 0.5$. What are its amplitude and phase responses?

6.56 The difference equation for a filter is: $y(nT) = 2y(nT - T) + 2x(nT) + x(nT - T)$. What are the locations of its poles and zeros?

6.57 What traditional filter type best describes the filter with the z transform:

$$H(z) = \frac{z^2 - 1}{z^2 + 1}$$

6.58 A discrete impulse function is applied to the inputs of four different filters. The output sequences of these filters are listed below. Which one of these filters has a pole outside the unit circle? (a) $\{1, 2, 3, 4, 5, 6, 0, 0, 0, \ldots\}$ (b) $\{1, -1, 1, -1, 1, -1, \ldots\}$ (c) $\{1, 2, 4, 8, 16, \ldots\}$ (d) $\{1, 0.5, 0.25, 0.125, \ldots\}$

6.59 Draw the block diagram of a filter that has the difference equation:

$$y(nT) = y(nT - T) + y(nT - 2T) + x(nT) + x(nT - T)$$

6.60 What is the transfer function $H(z)$ of a filter described by the difference equation:

$$y(nT) + 0.5y(nT - T) = x(nT)$$

6.61 A filter has an output sequence of $\{1, 5, 3, -9, 0, 0, \ldots\}$ in response to the input sequence of $\{1, 3, 0, 0, \ldots\}$. What is its transfer function?

7

Integer Filters

Jon D. Pfeffer

When digital filters must operate in a real-time environment, many filter designs become unsatisfactory due to the amount of required computation time. A considerable reduction in computation time is achieved by replacing floating-point coefficients with small integer coefficients in filter equations. This increases the speed at which the resulting filter program executes by replacing floating-point multiplication instructions with integer bit-shift and add instructions. These instructions require fewer clock cycles than floating-point calculations.

Integer filters are a special class of digital filters that have only integer coefficients in their defining equations. This characteristic leads to some design constraints that can often make it difficult to achieve features such as a sharp cutoff frequency. Since integer filters can operate at higher speeds than traditional designs, they are often the best type of filter for high sampling rates or when using a slow microprocessor.

This chapter discusses basic design concepts of integer filters. It also reviews why these filters display certain characteristics, such as linear phase, and how to derive the transfer functions. Next this theory is extended to show how to design low-pass, high-pass, bandpass, and band-reject integer filters by appropriate selection of pole and zero locations. Then a discussion shows how certain aspects of filter performance can be improved by cascading together integer filters. Next a recent design method is summarized which is more complicated but adds flexibility to the design constraints. Finally, a lab provides hands-on experience in the design and use of integer filters.

7.1 BASIC DESIGN CONCEPT

Lynn (1977) presented the best known techniques for integer filter design used today by showing a methodology to design low-pass, high-pass, bandpass, and band-reject filters with integer coefficients. This method is summarized in several steps. First place a number of evenly spaced zeros around the unit circle. These zeros completely attenuate the frequencies corresponding to their locations. Next choose poles that also lie on the unit circle to exactly cancel some of the zeros. When a pole cancels a zero, the frequency corresponding to this location is no longer attenuated. Since each point on the unit circle is representative of a frequency, the locations of the poles and zeros determine the frequency response of the filter. These filters are restricted types of recursive filters.

7.1.1 General form of transfer function

The general form of the filter transfer function used by Lynn for this application is:

$$H_1(z) = \frac{[1 - z^{-m}]^p}{[1 - 2\cos(\theta)z^{-1} + z^{-2}]^t} \tag{7.1a}$$

$$H_2(z) = \frac{[1 + z^{-m}]^p}{[1 - 2\cos(\theta)z^{-1} + z^{-2}]^t} \tag{7.1b}$$

The m exponent represents how many evenly spaced zeros to place around the unit circle. The angle θ represents the angular locations of the poles. The powers p and t represent the order of magnitude of the filter, which has a direct bearing on its gain and on the attenuation of the sidelobes. Raising p and t by equal integral amounts has the effect of cascading identical filters. If the filter is to be useful and physically realizable, p and t must both be nonnegative integers. The effects of raising p and t to values greater than one are further described in section 7.5.

7.1.2 Placement of poles

The denominator of this transfer function comes from the multiplication of a pair of complex-conjugate poles that always lie exactly on the unit circle. Euler's relation, $e^{j\theta} = \cos(\theta) + j\sin(\theta)$, shows that all values of $e^{j\theta}$ lie on the unit circle. Thus we can derive the denominator as follows:

$$\text{Denominator} = (z - e^{j\theta})(z - e^{-j\theta}) \tag{7.2a}$$

Multiplying the two factors, we get

$$\text{Denominator} = z^2 - (e^{j\theta} + e^{-j\theta})z + (e^{j\theta}e^{-j\theta}) \tag{7.2b}$$

Using the identity,

$$\cos(\theta) = \frac{e^{j\theta} + e^{-j\theta}}{2} \tag{7.2c}$$

We arrive at

$$\text{Denominator} = 1 - 2\cos(\theta)z^{-1} + z^{-2} \tag{7.2d}$$

The pair of complex-conjugate poles from the denominator of this transfer function provides integer multiplier and divisor coefficient values only when $2\cos(\theta)$ is an integer as shown in Figure 7.1(a). Such integer values result only when θ is equal to $0°$, $\pm 60°$, $\pm 90°$, $\pm 120°$, and $180°$ respectively, as shown in Figure 7.1(b). When the coefficients are integers, a multiplication step is saved in the calculations because the multiplication of small integers is done by using bit-shift C-language instructions instead of multiplication instructions.

The poles in these positions are guaranteed to exactly lie on the unit circle. It is impossible to "exactly" cancel a zero on the unit circle unless the pole has integer coefficients. All other locations, say $\theta = 15°$, have a small region of instability due to round-off error in the floating-point representation of numbers.

These filters have the added benefit of true-linear phase characteristics. This means that all the frequency components in the signal experience the same transmission delay through the filter (Lynn, 1972). This is important when it is desired to preserve the relative timing of the peaks and features of an output waveform. For example, it is crucial for a diagnostic quality ECG waveform to have P waves, T waves, and QRS complexes that remain unchanged in shape or timing. Changes in phase could reshape normal ECG waveforms to appear as possible arrhythmias, which is, of course, unacceptable.

7.1.3 Placement of zeros

We have seen that poles with integer coefficients that lie on the unit circle are restricted to five positions. This places constraints on the placement of zeros and also limits choices for the sampling rate when a designer wants to implement a specific cutoff frequency. To place zeros on the unit circle, one of two simple factors is used in the numerator of the filter's transfer function

$$(1 - z^{-m}) \tag{7.3a}$$

or

$$(1 + z^{-m}) \tag{7.3b}$$

For both equations, m is a positive integer equal to the number of zeros evenly spaced around the unit circle. Equation (7.3a) places a zero at $0°$ with the rest of the zeros evenly displaced by angles of $(360/m)°$. To prove this statement, set

$$(1 - z^{-m}) = 0$$

Placing the z term on the right-hand side of the equation

$$z^{-m} = 1$$

or equivalently

$$z^m = 1$$

You can see that, if z is on the unit circle, it is only equal to 1 at the point $z = (1, 0)$, which occurs every 2π radians. Thus, we can say

$$z^m = e^{jn2\pi} = 1$$

Solving for z, we arrive at

$$z = e^{j(n/m)2\pi}$$

Substituting in various integer values for n and m shows that the zeros are located at regular intervals on the unit circle every $(360/m)°$ beginning with a zero at $z = 1$. You can observe that the same solutions result when n is outside of the range $n = \{0, 1, 2, \dots, m-1\}$.

θ	$\cos(\theta)$	$2\cos(\theta)$
0°	1	2
±60°	+1/2	1
±90°	0	0
±120°	−1/2	−1
180°	−1	−2

(a)

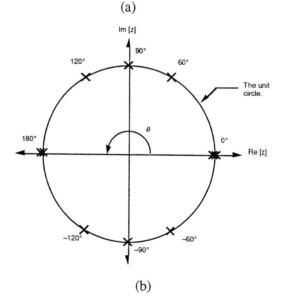

(b)

Figure 7.1 The possible pole placements given by the denominator of the transfer function in Eq. (7.1). (a) Table of only locations that result in integer coefficients. (b) Pole-zero plot corresponding to the table showing only possible pole locations on the unit circle. Notice that double poles are produced at 0° and 180°.

Equation (7.3b) also places m evenly spaced zeros around the unit circle every $(360/m)°$. Set

$$z^{-m} = -1$$

or equivalently

$$z^m = -1$$

We can solve for z in the same manner as described above. On the unit circle, $z = -1$ every $(2n + 1)\pi$ radians. The solution is

$$z = e^{j\,[(2n+1)/m]\pi}$$

where n and m again are integers.

Comparing the equations, $(1 - z^{-m})$ and $(1 + z^{-m})$, the difference in the placement of the zeros is a rotation of $1/2 \times (360/m)°$. Figure 7.2 shows a comparison of the results from each equation. When m is an odd number, $(1 + z^{-m})$ always places a zero at 180° and appears "flipped over" from the results of $(1 - z^{-m})$.

7.1.4 Stability and operational efficiency of design

For a filter to be stable, certain conditions must be satisfied. It is necessary to have the highest power of poles in the denominator be less than or equal to the number of zeros in the numerator. In other words, you cannot have fewer zeros than poles. This requirement is met for transfer functions of the form used in this chapter since they always have the same number of poles and zeros. Poles that are located at the origin do not show up in the denominator of the general transfer functions given in Eq. (7.1). You can more easily see these poles at the origin by manipulating their transfer functions to have all positive exponents. For example, if a low-pass filter of the type discussed in the next section has five zeros (i.e., $m = 5$), we can show that the denominator has at least five poles by multiplying the transfer function by one (i.e., z^5/z^5):

$$H(z) = \frac{1 - z^{-5}}{1 - z^{-1}} \times \frac{z^5}{z^5} = \frac{z^5 - 1}{z^4(z - 1)}$$

Poles at the origin simply have the effect of a time delay. In frequency domain terms, a time delay of m sampling periods is obtained by placing an mth-order pole at the origin of the z plane (Lynn, 1971).

A finite impulse response filter (FIR) has all its poles located at the origin. These are called trivial poles. Since we force the transfer function of Eq. (7.1) to have all its nontrivial poles exactly canceled by zeros, it is a FIR filter. This transfer function could be expressed without a denominator by not including the zeros and poles that exactly cancel out. For example

$$H(z) = \frac{1 - z^{-m}}{1 - z^{-1}} = 1 + z^1 + z^2 + z^3 + \ldots + z^{m+1} = \frac{Y(z)}{X(z)} \qquad (7.4)$$

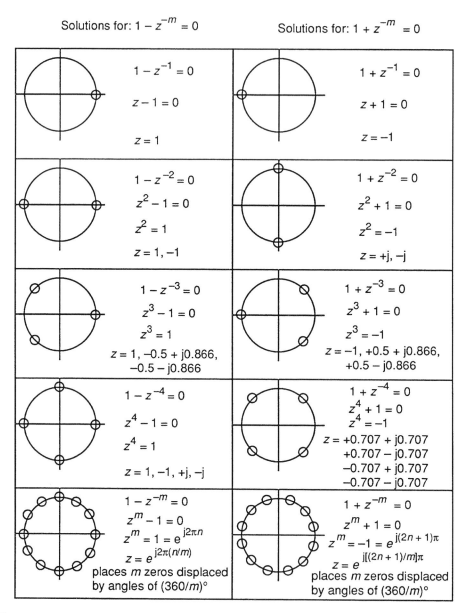

Solutions for: $1 - z^{-m} = 0$ Solutions for: $1 + z^{-m} = 0$

$1 - z^{-1} = 0$

$z - 1 = 0$

$z = 1$

$1 + z^{-1} = 0$

$z + 1 = 0$

$z = -1$

$1 - z^{-2} = 0$

$z^2 - 1 = 0$

$z^2 = 1$

$z = 1, -1$

$1 + z^{-2} = 0$

$z^2 + 1 = 0$

$z^2 = -1$

$z = +j, -j$

$1 - z^{-3} = 0$

$z^3 - 1 = 0$

$z^3 = 1$

$z = 1, -0.5 + j0.866,$
$-0.5 - j0.866$

$1 + z^{-3} = 0$

$z^3 + 1 = 0$

$z^3 = -1$

$z = -1, +0.5 + j0.866,$
$+0.5 - j0.866$

$1 - z^{-4} = 0$

$z^4 - 1 = 0$

$z^4 = 1$

$z = 1, -1, +j, -j$

$1 + z^{-4} = 0$

$z^4 + 1 = 0$
$z^4 = -1$

$z = +0.707 + j0.707$
$+0.707 - j0.707$
$-0.707 + j0.707$
$-0.707 - j0.707$

$1 - z^{-m} = 0$

$z^m - 1 = 0$

$z^m = 1 = e^{j2\pi n}$

$z = e^{j2\pi(n/m)}$

places m zeros displaced
by angles of $(360/m)°$

$1 + z^{-m} = 0$

$z^m + 1 = 0$

$z^m = -1 = e^{j(2n+1)\pi}$

$z = e^{j[(2n+1)/m]\pi}$

places m zeros displaced
by angles of $(360/m)°$

Figure 7.2 A comparison between the factors $(1 - z^{-m})$ and $(1 + z^{-m})$ for different values of m. When m is odd, the zeros appear "flipped over" from each other.

However, this is not done since expressing filters in recursive form is computationally more efficient. If m is large, the nonrecursive transfer function requires m additions in place of just one addition and one subtraction for the recursive method. In general, a transfer function can be expressed recursively with far fewer operations than in a nonrecursive form.

7.2 LOW-PASS INTEGER FILTERS

Using the transfer function of Eq. (7.1), we can design a low-pass filter by placing a pole at $z = (1, 0)$. However, the denominator produces two poles at $z = (1, 0)$ when $\cos(\theta) = 0°$. This problem is solved by either adding another zero at $z = (1, 0)$ or by removing a pole. The better solution is removing a pole since this creates a shorter, thus more efficient, transfer function. Also note from Figure 7.2 that the factor $(1 - z^{-m})$ should be used rather than $(1 + z^{-m})$ since it is necessary to have a zero positioned exactly at $0°$ on the unit circle. Thus, the transfer function for a recursive integer low-pass filter is

$$H(z) = \frac{1 - z^{-m}}{1 - z^{-1}} \tag{7.5}$$

This filter has a low-frequency lobe that is larger in magnitude than higher-frequency lobes; thus, it amplifies lower frequencies greater than the higher ones located in the smaller auxiliary sidelobes. These sidelobes result from the poles located at the origin of the z plane. Figure 7.3 shows the amplitude and phase responses for a filter with $m = 10$

$$H(z) = \frac{1 - z^{-10}}{1 - z^{-1}}$$

An intuitive feel for the amplitude response for all of the filters in this chapter is obtained by using the rubber membrane technique described Chapter 4. In short, when an extremely evenly elastic rubber membrane is stretched across the entire z plane, a zero "nails down" the membrane at its location. A pole stretches the membrane up to infinity at its location. Multiple poles at exactly the same location have the effect of stretching the membrane up to infinity, each additional one causing the membrane to stretch more tightly, thus driving it higher up at all locations around the pole. This has the effect of making a tent with a higher roof.

From this picture, we can infer how the amplitude response will look by equating it to the height of the membrane above the unit circle. The effective amplitude response is obtained from plotting the response from $\theta = 0°$ to $180°$. For angles greater than $180°$ the response is a reflection (or foldover) of the first $180°$. The digital filter should never receive frequencies in this range. They should have all been previously removed by an analog low-pass antialias filter.

To achieve lower cutoff frequencies, we must either add more zeros to the unit circle or use a lower sampling frequency. Adding more zeros is usually a better solution since it creates a sharper slope at the cutoff frequency and reduces the danger of information loss from low sampling rates. However, as the number of zeros increases, so does the gain of the filter. This can become a problem if a large output is not desired or if overflow errors occur.

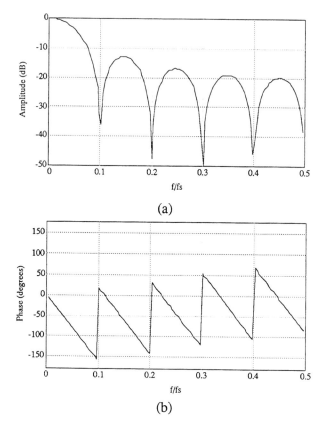

(a)

(b)

Figure 7.3 Low-pass filter with $m = 10$. (a) Amplitude response. (b) Phase response.

7.3 HIGH-PASS INTEGER FILTERS

There are several methods for designing a high-pass integer filter. Choice of an appropriate method depends on the desired cutoff frequency of the filter.

7.3.1 Standard high-pass filter design

Using the same design method described in section 7.2, a high-pass filter is constructed by placing a pole at the point $z = (-1, 0)$ corresponding to $\theta = 180°$ in the transfer function shown in section 7.1.1. For this to be possible, the numerator must also have a zero at the point $z = (-1, 0)$. This will always happen if the exponent m is an even number using the factor $(1 - z^{-m})$ in the numerator. A numerator using the factor $(1 + z^{-m})$ requires m to be an odd positive integer. As with the low-pass filter, the denominator of the general equation produces two poles at the location $z = (-1, 0)$, so one of these factors should be removed. The transfer function simplifies to one of two forms:

$$H(z) = \frac{1 - z^{-m}}{1 + z^{-1}} \qquad (m \text{ is even}) \qquad (7.6a)$$

or

$$H(z) = \frac{1 + z^{-m}}{1 + z^{-1}} \qquad (m \text{ is odd}) \qquad (7.6b)$$

The highest input frequency must be less than half the sampling frequency. This is not a problem since, in a good design, an analog antialias low-pass filter will eliminate all frequencies that are higher than one-half the sampling frequency. A high-pass filter with a cutoff frequency higher than half the sampling frequency is not physically realizable, thus making it useless.

To construct such a filter, zeros are placed on the unit circle and a pole cancels the zero at $\theta = 180°$. For a practical high-pass filter, the cutoff frequency must be greater than one-fourth the sampling frequency (i.e., $m \geq 4$). Increasing the number of zeros narrows the bandwidth. To achieve bandwidths greater than one-fourth the sampling frequency, requires the subtraction technique of the next section.

7.3.2 High-pass filter design based on filter subtraction

The frequency response of a composite filter formed by adding the outputs of two (or more) linear phase filters with the same transmission delay is equal to the simple algebraic sum of the individual responses (Ahlstrom and Tompkins, 1985). A high-pass filter $H_{high}(z)$ can also be designed by subtraction of a low-pass filter, $H_{low}(z)$ described in section 7.2, from an all-pass filter. This is shown graphically in Figure 7.4. An all-pass filter is a pure delay network with constant gain. Its transfer function can be represented as $H_a(z) = Az^{-m}$, where A is equal to the gain and m to number of zeros. Ideally, $H_a(z)$ should have the same gain as $H_{low}(z)$ at dc so that the difference is zero. Also the filter can operate more quickly if $A = 2^i$ where i is an integer so that a shift operation is used to scale the high-pass filter. To achieve minimal phase distortion, the number of delay units of the all-pass filter should be equal to the number of delay units needed to design the low-pass filter. Thus, we require that $\angle H_a(z) = \angle H_{low}(z)$. A low-pass filter with many zeros and thus a low cutoff frequency produces a high-pass filter with a low-frequency cutoff.

7.3.3 Special high-pass integer filters

We can also design high-pass filters for the special cases where every zero on the unit circle can potentially be canceled by a pole. This occurs when $m = 2, 4,$ or 6 with the factor $(1 - z^{-m})$. This concept can be extended for any even value of m in the factor $(1 - z^{-m})$ if noninteger coefficients are also included. If $m = 2$ and the zero at 180° is canceled, we have designed a high-pass filter with a nonsharp cutoff frequency starting at zero. If $m = 4$, we can either cancel the zeros with conjugate poles at 180° and 90°, or just at 180° if the filter should begin passing frequencies at one-fourth of the sampling rate. Similarly, if $m = 6$, zeros are canceled at 180°, 120°, and 60° to achieve similar results.

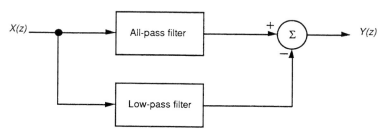

Figure 7.4 A block diagram of the low-pass filter being subtracted from a high-pass filter.

7.4 BANDPASS AND BAND-REJECT INTEGER FILTERS

In the design of an integer bandpass filter, once again one of the transfer functions of Eq. (7.1) is used. Unfortunately, the only possible choices of pole locations yielding integer coefficients are at 60°, 90°, and 120° [see Figure 7.1(a)]. The sampling frequency is chosen so that the passband frequency is at one of these three locations. Next a pair of complex-conjugate poles must be positioned so that one of them exactly cancels the zero in the passband frequency.

The choices for numerator are either $(1 - z^{-m})$ or $(1 + z^{-m})$. The best choice is the one that places a zero where it is needed with a reasonable value of m. The number of zeros chosen depends on the acceptable nominal bandwidth requirements. To get a very narrow bandwidth, we increase the number of zeros by using a higher power m. However, the gain of the filter increases with m, so the filter's output may become greater than the word size used, causing an overflow error. This is a severe error that must be avoided. As the number of zeros increases, (1) the bandwidth decreases, (2) the amplitudes of the neighboring sidelobes decrease, (3) the steepness of the cutoff increases, and (4) the difference equations require a greater history of the input data so that more sampled data values must be stored. Increasing the number of zeros of a bandpass filter increases the Q;

however, more ringing occurs in the filter's output. This is similar to the analog-equivalent filter. The effects of different values of Q are illustrated in section 12.5 for an ECG example.

The design of a band-reject (or bandstop) integer filter is achieved in one of two ways. The first solution is to simply place a zero on the unit circle at the frequency to be eliminated. The second method is to use the filter subtraction method to subtract a bandpass filter from an all-pass filter. The same restrictions and principles described in section 7.3.2 apply when using this method.

7.5 THE EFFECT OF FILTER CASCADES

The order of a filter is the number of identical filter stages that are used in series. The output of one filter provides the input to the next filter in the cascade. To increase the order of the general transfer function of Eq. (7.1), simply increase the exponents p and t by the same integer amount. For a filter of order n, the amplitude of the frequency response is expressed by $|H(z)|^n$. As n increases, the gain increases so you should be careful to avoid overflow error as mentioned in section 7.4. The gain of a filter can be attenuated by a factor of two by right bit-shifting the output values. However, this introduces a small quantization error as nonzero least-significant bits are shifted away.

All the filters previously discussed in this chapter suffer from substantial sidelobes and a poorly defined cutoff. The sidelobes can be substantially reduced by increasing the order of a filter by an even power. Figure 7.5 shows that, as the order of zeros in a filter increases, for example, *2nd* order, *4th* order, etc., the sidelobes become smaller and smaller.

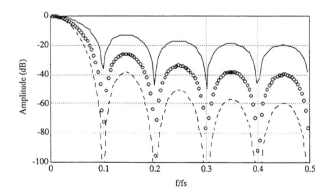

Figure 7.5 The effect of the order of a filter on its gain. This is an example of a high-pass filter with $m = 10$. Solid line: first order. Same as filter of Figure 7.3. Circles: second order. Dashed line: third order.

Increasing the filter order quickly reaches a point of diminishing returns as the filter recurrence formulas become more complex and the sidelobes are decreased by smaller increments.

The order of a filter can also be raised to an odd power; however, the phase response usually has better characteristics when the order is even. For the general transform of Eq. (7.1), the behavior of the phase response of filters of this type can be summarized as follows:

1. Even-order filters exhibit true-linear phase, thus eliminating phase distortion.
2. Odd-order filters have a piecewise-linear phase response. The discontinuities jump by 180° wherever the amplitude response is forced to zero by locating a zero on the unit circle.

A piecewise-linear phase response is acceptable if the filter displays significant attenuation in the regions where the phase response has changed to a new value. If the stopband is not well attenuated, phase distortions will occur whenever the signals that are being filtered have significant energy in those regions. The concept of phase distortions caused by linear phase FIR filters is thoroughly explained in a paper by Kohn (1987).

It is also possible to combine nonidentical filters together. Often complicated filters are crafted from simple subfilters. Examples of these include the high-pass and band-reject filters described in previous sections that were derived from filter subtraction. Lynn (1983) also makes use of this principle when he expands the design method described in this chapter to include resonator configurations. Lynn uses these resonator configurations to expand the design format to include 11 pole locations which add to design flexibility; however, this concept is not covered in this chapter.

7.6 OTHER FAST-OPERATING DESIGN TECHNIQUES

Principe and Smith (1986) used a method slightly different from Lynn's for designing digital filters to operate on electroencephalographic (EEG) data. Their method is a dual to the frequency sampling technique described in section 5.6. They construct the filter's transfer function in two steps. First zeros are placed on the unit circle creating several passbands, one of which corresponds to the desired passband. Next zeros are placed on the unit circle in the unwanted passbands to squelch gain and to produce stopbands.

An example of this is shown in Figure 7.6. Since all the poles are located at the origin, these FIR filters are guaranteed to be stable. Placing zeros to attenuate gain is more flexible than placing poles to cancel zeros on the unit circle. Zeros are placed in conjugate pairs by using the factor $1 - 2\cos(\theta) z^{-1} + z^{-2}$ in the numerator of the transfer function.

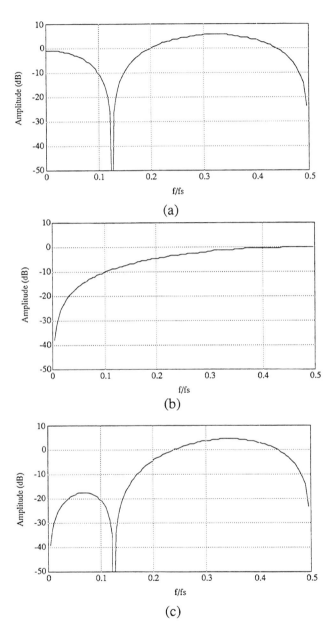

Figure 7.6 An example using only zeros to create a high-pass filter. The outputs of magnitude (a) plus magnitude (b) are summed to create magnitude (c).

If a zero is needed at a location that involves multiplication to calculate the coefficient, it is acceptable to place it at a nearby location that only requires a few binary-shift and add instructions so as to reduce computation time. A 3-bit approximation for $2\cos(\theta)$ which uses at most two shifts and two additions can place a zero at every location that the factors $(1 - z^{-m})$ or $(1 + z^{-m})$ do with an error of less than 6.5 percent if m is less than 8. This is a common technique to save multiplications. The zero can be slightly moved to an easy-to-calculate location at the cost of slightly decreasing the stopband attenuation.

This technique of squelching the stopband by adding more zeros could also be used with Lynn's method described in all of the previous sections. Adding zeros at specific locations can make it easier to achieve desired nominal bandwidths for bandpass filters, remove problem frequencies in the stopband (such as 60-Hz noise), or eliminate sidelobes without increasing the filter order.

All the previously discussed filters displayed a true-linear or piecewise-linear phase response. Sometimes situations demand a filter to have a sharp cutoff frequency, but phase distortion is irrelevant. For these cases, placing interior poles near zeros on the unit circle has the effect of amplifying the passband frequencies and attenuating the stopband frequencies. Whenever nontrivial poles remain uncanceled by zeros, an IIR rather than an FIR filter results. IIR filters have sharper cutoff slopes at the price of a nonlinear phase response. However, we do not wish to express the pole's coefficients, which have values between 0 and 1, using floating-point representation.

Thakor and Moreau (1987) solved this problem in a paper about the use of "quantized coefficients." To place poles inside the unit circle, they use a less restrictive method that retains some of the advantages of integer coefficients. They allow coefficients of the poles to have values of $x/2^y$ where x is an integer between 1 and $(2^y - 1)$ and y is a nonnegative integer called the quantization value. Quantization y is the designer's choice, but is limited by the microprocessor's word length (e.g., $y = 8$ for an 8-bit microprocessor). Using these values, fractional coefficients, such as 1/8, can be implemented with right shift instructions. A coefficient, such as 17/32, will not show much speed improvement over a multiplication instruction since many shifts and adds are required for its calculation. Thakor and Moreau give an excellent description of the use of these coefficients in filters and analyze the possible quantization, truncation, roundoff, filter-coefficient representation, and overflow errors that can occur.

7.7 DESIGN EXAMPLES AND TOOLS

This chapter includes enough material to design a wide variety of fast-operating digital filters. It would require many example problems to provide a feeling for all of the considerations needed to design an "optimal" filter for a certain application. However, the following design problems adequately demonstrate several of the methods used to theoretically analyze some of the characteristics of integer filters. We also demonstrate how a filter's difference equation is converted into C language so that it can be quickly implemented.

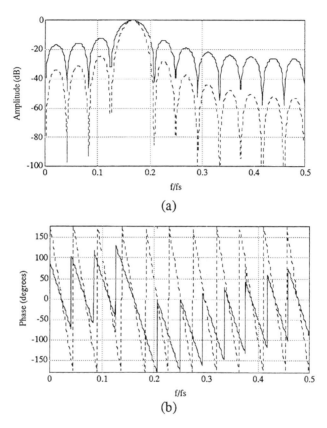

(a)

(b)

Figure 7.7 Piecewise-linear FIR bandpass filters. (a) Magnitude responses. Solid line: first-order filter with transfer function of $H(z) = (1 - z^{-24})/(1 - z^{-1} + z^{-2})$. Dashed line: second-order filter with transfer function of $H(z) = (1 - z^{-24})^2/(1 - z^{-1} + z^{-2})^2$. (b) Phase responses. Solid line: piecewise-linear phase response of first-order filter. Dashed line: true-linear phase response of second-order filter.

7.7.1 First-order and second-order bandpass filters

This section demonstrates the design of a bandpass filter that has 24 zeros evenly spaced around the unit circle beginning at 0°. The peak of the passband is located at 60°. A theoretical calculation of the amplitude and phase response is provided.

The transfer equation for this filter is:

$$H(z) = \frac{1 - z^{-24}}{1 - z^{-1} + z^{-2}} \tag{7.7}$$

Substitute $e^{j\omega T}$ for z and rearrange to produce positive and negative exponents of equal magnitude:

$$H(\omega T) = \frac{1 - e^{-j24\omega T}}{1 - e^{-j\omega T} + e^{-j2\omega T}} = \frac{e^{-j12\omega T} \, (e^{j12\omega T} - e^{-j12\omega T})}{e^{-j\omega T} \, (e^{j\omega T} - 1 + e^{-j\omega T})} \tag{7.8}$$

Substituting inside the parentheses the relation

$$e^{j\omega T} = \cos(\omega T) + j\sin(\omega T) \tag{7.9}$$

gives

$$H(\omega T) = \frac{e^{-j12\omega T} \, [\cos(12\omega T) + j\sin(12\omega T) - \cos(12\omega T) + j\sin(12\omega T)]}{e^{-j\omega T} \, [\cos(\omega T) + j\sin(\omega T) - 1 + \cos(\omega T) - j\sin(\omega T)]} \tag{7.10}$$

Combining terms

$$H(\omega T) = \frac{j2\sin(12\omega T)}{2\cos(\omega T) - 1} \times e^{-j11\omega T} \tag{7.11}$$

Substituting $j = e^{j\pi/2}$ gives

$$H(\omega T) = \frac{2\sin(12\omega T)}{2\cos(\omega T) - 1} \times e^{j(\pi/2 - 11\omega T)} \tag{7.12}$$

Thus, the magnitude response shown in Figure 7.7(a) as a solid line is

$$|H(\omega T)| = \left| \frac{2\sin(12\omega T)}{2\cos(\omega T) - 1} \right| \tag{7.13}$$

and the phase response shown in Figure 7.7(b) as a solid line is

$$\angle H(\omega T) = \pi/2 - 11\omega T \tag{7.14}$$

Next we cascade the filter with itself and calculate the amplitude and phase responses. This permits a comparison between the first-order and second-order filters.

The transfer equation now becomes

$$H(z) = \frac{(1 - z^{-24})^2}{(1 - z^{-1} + z^{-2})^2} \tag{7.15}$$

Again we substitute $e^{j\omega T}$ for z. The steps are similar to those for the first-order filter until we arrive at the squared form of Eq. (7.12). The transfer equation of the second-order example is

$$H(\omega T) = \left(\frac{2 \sin(12\omega T)}{2 \cos(\omega T) - 1} \times e^{j(\pi/2 - 11\omega T)} \right)^2$$

$$= \left(\frac{2\sin(12\omega T)}{2 \cos(\omega T) - 1} \right)^2 \times e^{j(\pi - 22\omega T)} \qquad (7.16)$$

The magnitude response of this second-order bandpass filter shown in Figure 7.7(a) as a dashed line is

$$|H(\omega T)| = \left| \left(\frac{2\sin(12\omega T)}{2 \cos(\omega T) - 1} \right)^2 \right| \qquad (7.17)$$

and the phase response shown in Figure 7.7(b) as a dashed line is

$$\angle H(\omega T) = \pi - 22\omega T \qquad (7.18)$$

Comparing the phase responses of the two filters, we see that the first-order filter is piecewise linear, whereas the second-order filter has a true-linear phase response. A true-linear phase response is recognized on these plots when every phase line has the same slope and travels from 360° to −360°.

To calculate the gain of the filter, substitute into the magnitude equation the critical frequency. For this bandpass filter, this frequency is located at an angle of 60°. Substituting this value into the magnitude response equation gives an indeterminate result

$$|H(\omega T)| = \left| \left(\frac{2\sin(12\omega T)}{2 \cos(\omega T) - 1} \right)^2 \right| = \left. \left| \left(\frac{2\sin(60°)}{2 \cos(60°) - 1} \right)^2 \right| \right|_{\omega T=60°} = \frac{3}{0} \quad (7.17)$$

Thus, to find the gain, L'Hôpital's Rule must be used. This method requires differentiation of the numerator and differentiation of the denominator prior to evaluation at 60°.

This procedure yields

$$\frac{d(|H(\omega T)|)}{d(\omega T)} = \left. \left| \left(\frac{24\cos(12\omega T)}{-2\sin(\omega T)} \right)^2 \right| \right|_{\omega T=60°} = \left| \left(\frac{24}{-\sqrt{3}} \right)^2 \right| = 192 \qquad (7.19)$$

Thus, the gain for the second-order filter is 192 compared to 13.9 for the first-order filter. Increasing the order of a filter also increases the filter's gain. However,

Figure 7.7(a) shows that the sidelobes are more attenuated than the passband lobe for the second-order filter.

We use the filter's difference equation to write a C-language function that implements this filter:

$$y(nT) = 2y(nT - T) - 3y(nT - 2T) + 2y(nT - 3T) - y(nT - 4T)$$
$$+ x(nT) - 2x(nT - 24T) + x(nT - 48T) \qquad (7.20)$$

Figure 7.8 shows the C-language program for a real-time implementation of the second-order bandpass filter. This code segment is straightforward; however, it is not the most time-efficient method for implementation. Each `for()` loop shifts all the values in each array one period in time.

```
/* Bandpass filter implementation of
 *
 *   H(z) = [( 1 - z^-24)^2] / [(1 - z^-1 + z^-2)^2]
 *
 * Notes: Static variables are automatically initialized to zero.
 * Their scope is local to the function. They retain their values
 * when the function is exited and reentered.
 *
 * Long integers are used for y, the output array. Since this
 * filter has a gain of 192, values for y can easily exceed the
 * values that a 16-bit integer can represent.
 */

long bandpass(int x_current)
{
register int i;
static int x[49];      /* history of input data points */
static long y[5];      /* history of difference equation outputs */

 for (i=4; i>0; i--)            /* shift past outputs */
    y[i] = y[i-1];

 for (i=48; i>0; i--)           /* shift past inputs */
    x[i] = x[i-1];

 x[0] = x_current;   /* update input array with new data point */

/* Implement difference equation */
 y[0] = 2*y[1] - 3*y[2] + 2*y[3] - y[4] + x[0] - 2*x[24] + x[48];

 return y[0];
}
```

Figure 7.8 C-language implementation of the second-order bandpass filter.

7.7.2 First-order low-pass filter

Here is a short C-language function for a six-zero low-pass filter. This function is passed a sample from the A/D converter. The data is filtered and returned. A FIFO circular buffer implements the unit delays by holding previous samples. This function consists of only one addition and one subtraction. It updates the pointer to the buffer rather than shifting all the data as in the previous example. This becomes more important as the number of zeros increases in the filter's difference equation. The difference equation for a six-zero low-pass filter is

$$y(nT) = y(nT - T) + x(nT) - x(nT - 6T) \qquad (7.21)$$

Figure 7.9 shows how to implement this equation in efficient C-language code.

```
/* Low-pass filter implementation of
 *
 *  H(z) = ( 1 - z^-6) / (1 - z^-1)
 *
 * Note 1: Static variables are initialized only once. Their scope
 * is local to the function. Unless set otherwise, they are
 * initialized to zero. They retain their values when the function
 * is exited and reentered.
 *
 * Note 2: This line increments pointer x_6delay along the x array
 * and wraps it to the first element when its location is at the
 * last element. The ++ must be a prefix to x_6delay; it is
 * equivalent to x_delay + 1.
 */

int lowpass(int x_current)
{
static x[6],              /* FIFO buffer of past samples */
       y,                 /* serves as both y(nT) and y(nT-T) */
       *x_6delay = &x[0}; /* pointer to x(nT-6T); see Note 1 */

y += (x_current - *x_6delay);   /* y(nT)=y(nT-T)+x(nT)-x(nT-6T) */
*x_6delay = x_current;       /* x_current becomes x(nT-T) in FIFO */
x_6delay = (x_6delay == &x[5]) ? &x[0] : ++x_6delay;
                                         /* See Note 2 */
 return(y);
}
```

Figure 7.9 Efficient C-language implementation of a first-order low-pass filter. Increments a pointer instead of shifting all the data.

7.8 LAB: INTEGER FILTERS FOR ECG ANALYSIS

Equipment designed for ECG analysis often must operate in real time. This means that every signal data point received by the instrument must be processed to produce an output before the next input data point is received. In the design process, cost constraints often make it desirable to use smaller, low-performance microprocessors to control a device. If the driver for an instrument is a PC, many times it is not equipped with a math coprocessor or a high-performance microprocessor. For this lab, you will design and implement several of the filters previously discussed to compare their performance in several situations.

Execute (F)ilters, (D)esign, then i(N)teger to enter the integer filter design shell.

1. Use the (G)enwave function to generate a normal ECG from template 1 and an abnormal ECG from template 5, both with a sampling rate of 100 sps and an amplitude resolution of eight bits.

(a) Design a bandpass filter for processing these signals with six zeros and a center frequency of 16.7 Hz. This type of filter is sometimes used in cardiotachometers to find the QRS complex determine heart rate.

(b) Filter the two ECGs with these filters. Since this filter has gain, you will probably need to adjust the amplitudes with the (Y) Sens function. Sketch the responses.

(c) Read the unit impulse signal file ups.dat from the STDLIB directory, and filter it. Sketch the response. Take the power spectrum of the impulse response using (P)wr Spect. The result is the amplitude response of the filter, which is the same as the response that you obtained when you designed the filter except that the frequency axis is double that of your design because the unit impulse signal was sampled at 200 sps instead of the 100 sps for which you designed your filter. Use the cursors of the (M)easure function to locate the 3-dB points, and find the bandwidth. Note that the actual bandwidth is half the measurement because of doubling of the frequency axis. Calculate the Q of the filter.

(d) Design two bandpass filters similar to the one in part (a) except with 18 and 48 zeros. Repeat parts (b) and (c) using these filters. As the number of zeros increases, the gain of these filters also increases, so you may have an overflow problem with some of the responses, particularly for the unit impulse. If this occurs, the output will appear to have discontinuities, indicating that the 16-bit representation of data in DigiScope has overflowed. In this case, attenuate the amplitude of the unit impulse signal prior to filtering it. Which of these three filter designs is most appropriate for detecting the QRS complexes? Why?

2. Design a low-pass filter with 10 zeros. Filter the normal ECG from part 1, and sketch the output. Which waves are attenuated and which are amplified? What is the 3-dB passband?

3. Generate a normal ECG with a sampling rate of 180 sps. Include 10% 60-Hz noise and 10% random noise in the signal. Design a single low-pass filter that

(a) has a low-pass bandwidth of about 15 Hz to attenuate random noise, and
(b) completely eliminates 60-Hz noise. Measure the actual 3-dB bandwidth.
Comment on the performance of your design.

7.9 REFERENCES

Ahlstrom, M. L., and Tompkins, W. J. 1985. Digital filters for real-time ECG signal processing
    using microprocessors. *IEEE Trans. Biomed. Eng.*, **BME-32**(9): 708–13.
Kohn, A. F. 1987. Phase distortion in biological signal analysis caused by linear phase FIR
    filters. *Med. & Biol. Eng. & Comput.* 25: 231–38.
Lynn, P. A. 1977. Online digital filters for biological signals: some fast designs for a small
    computer. *Med. & Biol. Eng. & Comput.* 15: 534–40.
Lynn, P. A. 1971. Recursive digital filters for biological signals. *Med. & Biol. Eng. & Comput.*
    9: 37–44.
Lynn, P. A. 1972. Recursive digital filters with linear-phase characteristics. *Comput. J.* 15: 337–
    42.
Lynn, P. A. 1983. Transversal resonator digital filters: fast and flexible online processors for
    biological signals. *Med. & Biol. Eng. & Comput.* 21: 718–30.
Principe, J. C., and Smith, J. R. 1986. Design and implementation of linear phase FIR filters for
    biological signal processing. *IEEE Trans. Biomed. Eng.* **BME-33**(6):550–59.
Thakor, N. V., and Moreau, D. 1987. Design and analysis of quantised coefficient digital filters:
    application to biomedical signal processing with microprocessors. *Med. & Biol. & Comput.*
    25: 18–25.

7.10 STUDY QUESTIONS

7.1   Why is it advantageous to use integer coefficients in a digital filter's difference equation?
7.2   Explain why it is more difficult to design a digital filter when all coefficients are restricted
      to having integer values.
7.3   Show how the denominator of Eq. (7.1) will always place two poles on the unit circle for
      all values of θ. What values of θ produce integer coefficients? Why should values of θ that
      yield floating-point numbers be avoided?
7.4   Show mathematically why the numerators $(1 + z^{-m})$ and $(1 - z^{-m})$ place zeros on the unit
      circle. Both numerators produce evenly spaced zeros. When would it be advantageous to
      use $(1 - z^{-m})$ instead of $(1 + z^{-m})$?
7.5   When does Eq. (7.1) behave as a FIR filter? How can this equation become unstable? What
      are the advantages of expressing a FIR filter in recursive form?
7.6   When does Eq. (7.1) behave as a low-pass filter? Discuss what characteristics of filter
      behavior a designer must consider when a filter is to have a low cutoff frequency.
7.7   What is the difference between a true-linear phase response and a piecewise-linear phase
      response? When is a linear phase response essential? Can a filter with a piecewise-linear
      phase response behave as one with a true-linear phase response?
7.8   Name four ways in which an integer digital filter's magnitude and phase response change
      when the filter is cascaded with itself. Why or why not are these changes helpful?
7.9   If poles and zeros are placed at noninteger locations, how can a digital filter still remain
      computationally efficient? Describe two methods that use this principle.

7.10 Calculate expressions for the amplitude and phase response of a filter with the z transform

$$H(z) = 1 - z^{-6}$$

7.11 The numerator of a transfer function is $(1 - z^{-10})$. Where are its zeros located?

7.12 A filter has 12 zeros located on the unit circle starting at dc and equally spaced at $30°$ increments (i.e., $1 - z^{-12}$). There are three poles located at $z = +0.9$, and $z = \pm j$. The sampling frequency is 360 samples/s. (a) At what frequency is the output at its maximal amplitude? (b) What is the gain at this frequency?

7.13 A digital filter has the following transfer function. (a) What traditional filter type best describes this filter? (b) What is its gain at dc?

$$H(z) = \frac{1 - z^{-6}}{(1 - z^{-1})(1 - z^{-1} + z^{-2})}$$

7.14 For a filter with the following transfer function, what is the (a) amplitude response, (b) phase response, (c) difference equation?

$$H(z) = \frac{1 - z^{-8}}{1 + z^{-2}}$$

7.15 A digital filter has the following transfer function. (a) What traditional filter type best describes this filter? (b) Draw its pole-zero plot. (c) Calculate its amplitude response. (d) What is its difference equation?

$$H(z) = \frac{(1 - z^{-8})^2}{(1 + z^{-2})^2}$$

7.16 What is the gain of a filter with the transfer function

$$H(z) = \frac{1 - z^{-6}}{1 - z^{-1}}$$

7.17 What traditional filter type best describes a filter with the transfer function

$$H(z) = \frac{1 - z^{-256}}{1 - z^{-128}}$$

7.18 What traditional filter type best describes a filter with the transfer function

$$H(z) = \frac{1 - z^{-200}}{1 - z^{-2}}$$

7.19 A digital filter has four zeros located at $z = \pm 1$ and $z = \pm j$ and four poles located at $z = 0$, $z = 0$, and $z = \pm j$. The sampling frequency is 800 samples/s. The maximal output amplitude occurs at what frequency?

7.20 For a sampling rate of 100 samples/s, a digital filter with the following transfer function has its maximal gain at approximately what frequency (in Hz)?

$$H(z) = \frac{1 - z^{-36}}{1 - z^{-1} + z^{-2}}$$

7.21 The z transform of a filter is:

$$H(z) = 1 - z^{-360}$$

The following sine wave is applied at the input: $x(t) = 100 \sin(2\pi 10t)$. The sampling rate is 720 samples/s. (a) What is the peak-to-peak output of the filter? (b) If a *unit step* input is applied, what will the output amplitude be after 361 samples? (c) Where could poles be placed to convert this to a bandpass filter with integer coefficients?

7.22 What is the phase delay (in milliseconds) through the following filter which operates at 200 samples/sec?

$$H(z) = \frac{1 - z^{-100}}{1 - z^{-2}}$$

7.23 A filter has 8 zeros located on the unit circle starting at dc and equally spaced at 45° increments. There are two poles located at $z = \pm j$. The sampling frequency is 360 samples/s. What is the gain of the filter?

8

Adaptive Filters

Steven Tang

This chapter discusses how to build adaptive digital filters to perform noise cancellation and signal extraction. Adaptive techniques are advantageous because they do not require a priori knowledge of the signal or noise characteristics as do fixed filters. Adaptive filters employ a method of learning through an estimated synthesis of a desired signal and error feedback to modify the filter parameters. Adaptive techniques have been used in filtering of 60-Hz line frequency noise from ECG signals, extracting fetal ECG signals, and enhancing P waves, as well as for removing other artifacts from the ECG signal. This chapter provides the basic principles of adaptive digital filtering and demonstrates some direct applications.

In digital signal processing applications, frequently a desired signal is corrupted by interfering noise. In fixed filter methods, the basic premise behind optimal filtering is that we must have knowledge of both the signal and noise characteristics. It is also generally assumed that the statistics of both sources are well behaved or wide-sense stationary. An adaptive filter learns the statistics of the input sources and tracks them if they vary slowly.

8.1 PRINCIPAL NOISE CANCELER MODEL

In biomedical signal processing, adaptive techniques are valuable for eliminating noise interference. Figure 8.1 shows a general model of an adaptive filter noise canceler. In the discrete time case, we can model the primary input as $s(nT) + n_0(nT)$. The noise is additive and considered uncorrelated with the signal source. A secondary reference input to the filter feeds a noise $n_1(nT)$ into the filter to produce output $\zeta(nT)$ that is a close estimate of $n_0(nT)$. The noise $n_1(nT)$ is correlated in an unknown way to $n_0(nT)$.

The output $\zeta(nT)$ is subtracted from the primary input to produce the system output $y(nT)$. This output is also the error $\varepsilon(nT)$ that is used to adjust the taps of the adaptive filter coefficients $\{w(1,\dots,p)\}$.

$$y(nT) = s(nT) + n_0(nT) - \zeta(nT) \qquad (8.1)$$

Squaring the output and making the (nT) implicit to simplify each term

$$y^2 = s^2 + (n_0 - \zeta)^2 + 2s(n_0 - \zeta) \qquad (8.2)$$

Taking the expectation of both sides,

$$E[y^2] = E[s^2] + E[(n_0 - \zeta)^2] + 2E[s(n_0 - \zeta)]$$

$$= E[s^2] + E[(n_0 - \zeta)^2] \qquad (8.3)$$

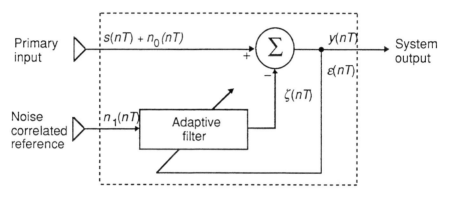

Figure 8.1 The structure of an adaptive filter noise canceler.

Since the signal power $E[s^2]$ is unaffected by adjustments to the filter

$$\min E[y^2] = E[s^2] + \min E[(n_0 - \zeta)^2] \qquad (8.4)$$

When the system output power is minimized according to Eq. (8.4), the mean-squared error (MSE) of $(n_0 - \zeta)$ is minimum, and the filter has adaptively learned to synthesize the noise $(\zeta \approx n_0)$. This approach of iteratively modifying the filter coefficients using the MSE is called the Least Mean Squared (LMS) algorithm.

8.2 60-HZ ADAPTIVE CANCELING USING A SINE WAVE MODEL

It is well documented that ECG amplifiers are corrupted by a sinusoidal 60-Hz line frequency noise (Huhta and Webster, 1973). As discussed in Chapter 5, a non-recursive band-reject notch filter can be implemented to reduce the power of noise at 60 Hz. The drawbacks to this design are that, while output noise power is reduced, such a filter (1) also removes the 60-Hz component of the signal, (2) has a very slow rolloff that unnecessarily attenuates other frequency bands, and (3) becomes nonoptimal if either the amplitude or the frequency characteristics of the noise change. Adaptive transversal (tapped delay line) filters allow for elimination of noise while maintaining an optimal signal-to-noise ratio for nonstationary processes.

One simplified method for removal of 60-Hz noise is to model the reference source as a 60-Hz sine wave (Ahlstrom and Tompkins, 1985). The only adaptive parameter is the amplitude of the sine wave. Figure 8.2 shows three signals: $x(nT)$ is the input ECG signal corrupted with 60-Hz noise, $e(nT)$ is the estimation of the noise using a 60-Hz sine wave, and $y(nT)$ is the output of the filter.

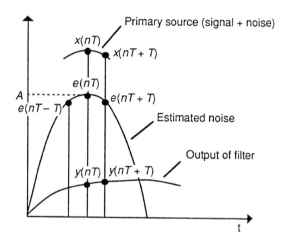

Figure 8.2 Sine wave model for 60-Hz adaptive cancellation.

The algorithm begins by estimating the noise as an assumed sinusoid with amplitude A and frequency ω

$$e(nT) = A\sin(\omega nT) \qquad (8.5)$$

In this equation, we replace term (nT) by $(nT - T)$ to find an expression for the estimated signal one period in the past. This substitution gives

$$e(nT - T) = A\sin(\omega nT - \omega T) \qquad (8.6)$$

Similarly, an expression that estimates the next point in the future is obtained by replacing (nT) by $(nT + T)$ in Eq. (8.5), giving

$$e(nT + T) = A\sin(\omega nT + \omega T) \qquad (8.7)$$

We now recall a trigonometric identity

$$\sin(\alpha + \beta) = 2\sin(\alpha)\cos(\beta) - \sin(\alpha - \beta) \qquad (8.8)$$

Now let

$$\alpha = \omega nT \quad and \quad \beta = \omega T \qquad (8.9)$$

Expanding the estimate for the future estimate of Eq. (8.7) using Eqs. (8.8) and (8.9) gives

$$e(nT + T) = 2\,A\sin(\omega nT)\ \cos(\omega T) - A\sin(\omega nT - \omega T) \qquad (8.10)$$

Note that the first underlined term is the same as the expression for $e(nT)$ in Eq. (8.5), and the second underlined term is the same as the expression for $e(nT - T)$ in Eq. (8.6). The term, $\cos(\omega T)$, is a constant determined by the frequency of the noise ω to be eliminated and by the sampling frequency, $f_S = 1/T$:

$$N = \cos(\omega T) = \cos\left(\frac{2\pi f}{f_S}\right) \qquad (8.11)$$

Thus, Eq. (8.10) is rewritten, giving a relation for the future estimated point on a sampled sinusoidal noise waveform based on the values at the current and past sample times.

$$e(nT + T) = 2Ne(nT) - e(nT - T) \qquad (8.12)$$

The output of the filter is the difference between the input and the estimated signals

$$y(nT + T) = x(nT + T) - e(nT + T) \qquad (8.13)$$

Thus, if the input were only noise and the estimate were exactly tracking (i.e., modeling) it, the output would be zero. If an ECG were superimposed on the input noise, it would appear noise-free at the output.

The ECG signal is actually treated as a transient, while the filter iteratively attempts to change the "weight" or amplitude of the reference input to match the desired signal, the 60-Hz noise. The filter essentially learns the amount of noise that is present in the primary input and subtracts it out. In order to iteratively adjust the filter to adapt to changes in the noise signal, we need feedback to adjust the sinusoidal amplitude of the estimate signal for each sample period.

We define the difference function

$$f(nT + T) = [x(nT + T) - e(nT + T)] - [x(nT) - e(nT)] \qquad (8.14)$$

In order to understand this function, consider Figure 8.3. Our original model of the noise $e(nT)$ in Eq. (8.5) assumed a simple sine wave with no dc component as shown. Typically, however, there is a dc offset represented by V_{dc} in the input $x(nT)$ signal. From the figure

$$V_{dc}(nT + T) = x(nT + T) - e(nT + T) \qquad (8.15)$$

and also

$$V_{dc}(nT) = x(nT) - e(nT) \qquad (8.16)$$

Assuming that the dc level does not change significantly between samples, then

$$V_{dc}(nT + T) - V_{dc}(nT) = 0 \qquad (8.17)$$

This subtraction of the terms representing the dc level in Eqs. (8.15) and (8.16) is the basis for the function in Eq. (8.14). It subtracts the dc while simultaneously comparing the input and estimated waveforms.

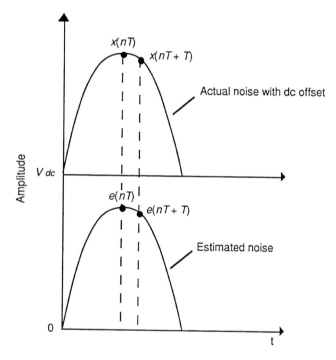

Figure 8.3 The actual noise waveform may include a dc offset that was included in the original model of the estimated signal.

We use $f(nT + T)$ to determine if the estimate $e(nT)$ was too large or too small. If $f(nT + T) = 0$, the estimate is correct and there is no need to adjust the future estimate, or

$$e(nT + T) = e(nT + T) \tag{8.18}$$

If $f(nT + T) > 0$, the estimate is low, and the estimate is adjusted upward by a small step size d

$$e(nT + T) = e(nT + T) + d \tag{8.19}$$

If $f(nT + T) < 0$, the estimate is high and the estimate is adjusted downward by a small step size d

$$e(nT + T) = e(nT + T) - d \tag{8.20}$$

The choice of d is empirically determined and depends on how quickly the filter needs to adapt to changes in the interfering noise. If d is large, then the filter quickly adjusts its coefficients after the onset of 60-Hz noise. However, if d is too large, the filter will not be able to converge exactly to the noise. This results in small oscillations in the estimated signal once the correct amplitude has been found. With a smaller d, the filter requires a longer learning period but provides more exact tracking of the noise for a smoother output. If the value of d is too large or too small, the filter will never converge to a proper noise estimate.

A typical value of d is less than the least significant bit value of the integers used to represent a signal. For example, if the full range of numbers from an 8-bit A/D converter is 0–255, then an optimal value for d might be 1/4.

Producing the estimated signal of Eq. (8.12) requires multiplication by a fraction N given in Eq. (8.11). For a sampling rate of 500 sps and 60-Hz power line noise

$$N = \cos\left(\frac{2\pi \times 60}{500}\right) = 0.7289686 \tag{8.21}$$

Such a multiplier requires floating-point arithmetic, which could considerably slow down the algorithm. In order to approximate such a multiplier, we might choose to use a summation of power-of-two fractions, which could be implemented with bit-shift operations and may be faster than floating-point multiplication in some hardware environments. In this case

$$N = \frac{1}{2} + \frac{1}{8} + \frac{1}{16} + \frac{1}{32} + \frac{1}{128} + \frac{1}{512} + \frac{1}{2048} = 0.72900 \tag{8.22}$$

8.3 OTHER APPLICATIONS OF ADAPTIVE FILTERING

Adaptive filtering is not only used to suppress 60-Hz interference but also for signal extraction and artifact cancellation. The adaptive technique is advantageous for generating a desired signal from one that is uncorrelated with it.

8.3.1 Maternal ECG in fetal ECG

Prenatal monitoring has made it possible to detect the heartbeat of the unborn child noninvasively. However, motion artifact and the maternal ECG make it very difficult to perceive the fetal ECG since it is a low-amplitude signal. Adaptive filtering has been used to eliminate the maternal ECG. Zhou et al. (1985) describe an algorithm that uses a windowed LMS routine to adapt the tap weights. The abdominal lead serves as the primary input and the chest lead from the mother is used as the reference noise input. Subtracting the best matched maternal ECG from the abdominal ECG which contains both the fetal and maternal ECGs produces a residual signal that is the fetal ECG.

8.3.2 Cardiogenic artifact

The area of electrical impedance pneumography has used adaptive filtering to solve the problem of cardiogenic artifact (ZCG). Such artifact can arise from electrical impedance changes due to blood flow and heart-volume changes. This can lead to a false interpretation of breathing. When monitoring for infant apnea, this might result in a failure to alarm. Sahakian and Kuo (1985) proposed using an adaptive LMS algorithm to extract the cardiogenic impedance component so as to achieve the best estimate of the respiratory impedance component. To model the cardiogenic artifact, they created a template synchronized to the QRS complex in an ECG that included sinus arrhythmia. Cardiogenic artifact is synchronous with but delayed from ventricular systole, so the ECG template can be used to derive and eliminate the ZCG.

8.3.3 Detection of ventricular fibrillation and tachycardia

Ventricular fibrillation detection has generally used frequency-domain techniques. This is computationally expensive and cannot always be implemented in real time. Hamilton and Tompkins (1987) describe a unique method of adaptive filtering to locate the poles corresponding to the frequency spectrum formants. By running a second-order IIR filter, the poles derived from the coefficients give a fairly good estimate of the first frequency peak.

The corresponding z transform of such a filter is

$$H(z) = \frac{1}{1 - b_1 z^{-1} - b_2 z^{-2}}$$

We can solve for the pole radius and angle by noting that, for all poles not on the real axis

$$b_1 = 2r\cos\theta \quad \text{and} \quad b_2 = -r^2$$

Using the fact that fibrillation produces a prominent peak in the 3–7 Hz frequency band, we can determine whether the poles fall in the "detection region" of the z plane. An LMS algorithm updates the coefficients of the filter. Figure 8.4 shows the z-plane pole-zero diagram of the adaptive filter. The shaded region indicates that the primary peak in the frequency spectrum of the ECG is in a "dangerous" area. The only weakness of the algorithm is that it creates false detections for rhythm rates greater than 100 bpm with frequent PVCs, atrial fibrillation, and severe motion artifact.

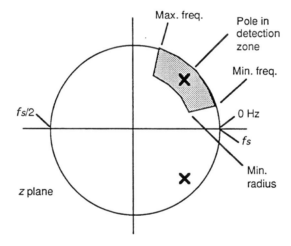

Figure 8.4 The z plane showing the complex-conjugate poles of the second-order adaptive filter.

8.4 LAB: 60-HZ ADAPTIVE FILTER

Load UW DigiScope, select **ad(V) Ops**, then **(A)daptive**. This module is a demonstration of a 60-Hz canceling adaptive filter as described in the text. You have control over the filter's step size d. This controls how quickly the filter *learns* the amount of 60 Hz in the signal. By turning the 60-Hz noise off after the filter has adapted out the noise, you can observe that the filter must now unlearn the 60-Hz component. This routine always uses the same data file `adapting.dat` to which the 60-Hz noise is added.

8.5 REFERENCES

Ahlstrom, M. L., and Tompkins, W. J. 1985. Digital filters for real-time ECG signal processing using microprocessors. *IEEE Trans. Biomed. Eng.,* **BME-32**(9): 708–13.
Hamilton, P. S., and Tompkins, W. J. 1987. Detection of ventricular fibrillation and tachycardia by adaptive modeling. *Proc. Annu. Conf. Eng. Med. Bio. Soc.,* 1881–82.
Haykin, S. 1986. *Adaptive Filter Theory.* Englewood Cliffs, NJ: Prentice Hall.
Huhta, J. C., and Webster, J. G. 1973. 60-Hz interference in electrocardiography. *IEEE Trans. Biomed. Eng.,* **BME-20**(2): 91-101.
Mortara, D. W. 1977. Digital filters for ECG signals. *Computers in Cardiology,* 511–14.
Sahakian, A. V., and Kuo, K. H. 1985. Canceling the cardiogenic artifact in impedance pneumography. *Proc. Annu. Conf. Eng. Med. Bio. Soc.,* 855–59.
Sheng, Z. and Thakor, N. V. 1987. P-wave detection by an adaptive QRS-T cancellation technique. *Computers in Cardiology,* 249–52.
Widrow, B., Glover, J. R., John, M., Kaunitz, J., Charles, S. J., Hearn, R. H., Zeidler, J. R., Dong, E., and Goodlin, R. C. 1975. Adaptive noise canceling: principles and applications. *Proc. IEEE,* **63**(12): 1692–1716.
Zhou, L., Wei, D., and Sun, L. 1985. Fetal ECG processing by adaptive noise cancellation. *Proc. Annu. Conf. Eng. Med. Bio. Soc.,* 834–37.

8.6 STUDY QUESTIONS

8.1 What are the main advantages of adaptive filters over fixed filters?
8.2 Explain the criterion that is used to construct a Wiener filter.
8.3 Why is the error residual of a Wiener filter normal to the output?
8.4 Design an adaptive filter using the method of steepest-descent.
8.5 Design an adaptive filter using the LMS algorithm.
8.6 Why are bounds necessary on the step size of the steepest-descent and LMS algorithms?
8.7 What are the costs and benefits of using different step sizes in the 60-Hz sine wave algorithm?
8.8 Explain how the 60-Hz sine wave algorithm adapts to the phase of the noise.
8.9 The adaptive 60-Hz filter calculates a function

$$f(nT + T) = [x(nT + T) - e(nT + T)] - [x(nT) - e(nT)]$$

If this function is less than zero, how does the algorithm adjust the future estimate, $e(nT + T)$?

8.10 The adaptive 60-Hz filter uses the following equation to estimate the noise:

$$e(nT + T) = 2Ne(nT) - e(nT - T)$$

If the future estimate is found to be too high, what adjustment is made to (a) $e(nT - T)$, (b) $e(nT + T)$. (c) Write the equation for N and explain the terms of the equation.

8.11 The adaptive 60-Hz filter calculates the function

$$f(nT + T) = [x(nT + T) - e(nT + T)] - [x(nT) - e(nT)]$$

It adjusts the future estimate $e(nT + T)$ based on whether this function is greater than, less than, or equal to zero. Use a drawing and explain why the function could not be simplified to

$$f(nT + T) = x(nT + T) - e(nT + T)$$

9

Signal Averaging

Pradeep Tagare

Linear digital filters like those discussed in previous chapters perform very well when the spectra of the signal and noise do not significantly overlap. For example, a low-pass filter with a cutoff frequency of 100 Hz generally works well for attenuating noise frequencies greater than 100 Hz in ECG signals. However, if high-level noise frequencies were to span the frequency range from 50–100 Hz, attempting to remove them using a 50–Hz low-pass filter would attenuate some of the components of the ECG signal as well as the noise. High-amplitude noise corruption within the frequency band of the signal may completely obscure the signal. Thus, conventional filtering schemes fail when the signal and noise frequency spectra significantly overlap. Signal averaging is a digital technique for separating a repetitive signal from noise without introducing signal distortion (Tompkins and Webster, 1981). This chapter describes the technique of signal averaging for increasing the signal-to-noise ratio and discusses several applications.

9.1 BASICS OF SIGNAL AVERAGING

Figure 9.1(a) shows the spectrum of a signal that is corrupted by noise. In this case, the noise bandwidth is completely separated from the signal bandwidth, so the noise can easily be discarded by applying a linear low-pass filter. On the other hand, the noise bandwidth in Figure 9.1(b) overlaps the signal bandwidth, and the noise amplitude is larger than the signal. For this situation, a low-pass filter would need to discard some of the signal energy in order to remove the noise, thereby distorting the signal.

One predominant application area of signal averaging is in electroencephalography. The EEG recorded from scalp electrodes is difficult to interpret in part because it consists of a summation of the activity of the billions of brain cells. It is impossible to deduce much about the activity of the visual or auditory parts of the brain from the EEG. However, if we stimulate a part of the brain with a flash of light or an acoustical click, an evoked response occurs in the region of the brain that processes information for the sensory system being stimulated. By summing the signals that are evoked immediately following many stimuli and dividing by the total number of stimuli, we obtain an averaged evoked response. This signal can reveal a great deal about the performance of a sensory system.

Signal averaging sums a set of time epochs of the signal together with the super-imposed random noise. If the time epochs are properly aligned, the signal wave-forms directly sum together. On the other hand, the uncorrelated noise averages out in time. Thus, the signal-to-noise ratio (SNR) is improved.

Signal averaging is based on the following characteristics of the signal and the noise:

1. The signal waveform must be repetitive (although it does not have to be peri-odic). This means that the signal must occur more than once but not necessarily at regular intervals.
2. The noise must be random and uncorrelated with the signal. In this application, random means that the noise is not periodic and that it can only be described statistically (e.g., by its mean and variance).
3. The temporal position of each signal waveform must be accurately known.

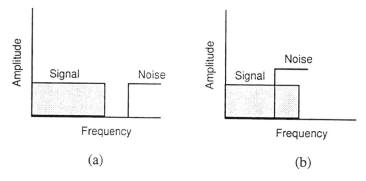

Figure 9.1 Signal and noise spectra. (a) The signal and noise bands do not overlap, so a conven-tional low-pass filter can be used to retain the signal and discard the noise. (b) Since the signal and noise spectra overlap, conventional filters cannot be used to discard the noise frequencies without discarding some signal energy. Signal averaging may be useful in this case.

It is the random nature of noise that makes signal averaging useful. Each time epoch (or sweep) is intentionally aligned with the previous epochs so that the digi-tized samples from the new epoch are added to the corresponding samples from the previous epochs. Thus the time-aligned repetitive signals S in each epoch are added directly together so that after four epochs, the signal amplitude is four times larger than for one epoch ($4S$). If the noise is random and has a mean of zero and an aver-age rms value N, the rms value after four epochs is the square root of the sum of squares (i.e., $(4N^2)^{1/2}$ or $2N$). In general after m repetitions the signal amplitude is

mS and the noise amplitude is $(m)^{1/2}N$. Thus, the SNR improves as the ratio of m to $m^{1/2}$ (i.e., $m^{1/2}$). For example, averaging 100 repetitions of a signal improves the SNR by a factor of 10. This can be proven mathematically as follows.

The input waveform $f(t)$ has a signal portion $S(t)$ and a noise portion $N(t)$. Then

$$f(t) = S(t) + N(t) \tag{9.1}$$

Let $f(t)$ be sampled every T seconds. The value of any sample point in the time epoch ($i = 1, 2,..., n$) is the sum of the noise component and the signal component.

$$f(iT) = S(iT) + N(iT) \tag{9.2}$$

Each sample point is stored in memory. The value stored in memory location i after m repetitions is

$$\sum_{k=1}^{m} f(iT) = \sum_{k=1}^{m} S(iT) + \sum_{k=1}^{m} N(iT) \quad \text{for} \quad i = 1, 2, ..., n \tag{9.3}$$

The signal component for sample point i is the same at each repetition if the signal is stable and the sweeps are aligned together perfectly. Then

$$\sum_{k=1}^{m} S(iT) = mS(iT) \tag{9.4}$$

The assumptions for this development are that the signal and noise are uncorrelated and that the noise is random with a mean of zero. After many repetitions, $N(iT)$ has an rms value of σ_n.

$$\sum_{k=1}^{m} N(iT) = \sqrt{m\sigma_n^2} = \sqrt{m}\,\sigma_n \tag{9.5}$$

Taking the ratio of Eqs. (9.4) and (9.5) gives the SNR after m repetitions as

$$\text{SNR}_m = \frac{mS(iT)}{\sqrt{m}\,\sigma_n} = \sqrt{m}\,\text{SNR} \tag{9.6}$$

Thus, signal averaging improves the SNR by a factor of $\sqrt{m}$. Figure 9.2 is a graph illustrating the results of Eq. (9.6).

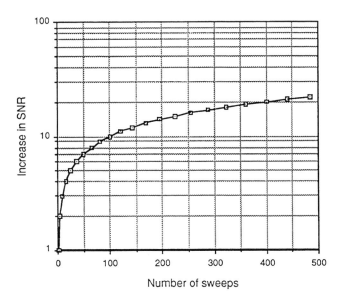

Figure 9.2 Increase in SNR as a function of the number of sweeps averaged.

Figure 9.3 illustrates the problem of signal averaging. The top trace is the ECG of the middle trace after being corrupted by random noise. Since the noise is broadband, there is no way to completely remove it with a traditional linear filter without also removing some of the ECG frequency components, thereby distorting the ECG. Signal averaging of this noisy signal requires a way to time align each of the QRS complexes with the others. By analyzing a heavily filtered version of the waveform, it is possible to locate the peaks of the QRS complexes and use them for time alignment. The lower trace shows these timing references (fiducial points) that are required for signal processing.

Figure 9.4 shows how the QRS complexes, centered on the fiducial points, are assembled and summed to produce the averaged signal. The time-aligned QRS complexes sum directly while the noise averages out to zero. The fiducial marks may also be located before or after the signal to be averaged, as long as they have accurate temporal relationships to the signals.

One research area in electrocardiography is the study of late potentials that require an ECG amplifier with a bandwidth of 500 Hz. These small, high-frequency signals of possible clinical significance occur after the QRS complex in body surface ECGs of abnormals. These signals are so small compared to the other waveforms in the ECG that they are hidden in the noise and are not observable without signal averaging. In this application, the fiducial points are derived from the QRS complexes, and the averaging region is the time following each QRS complex.

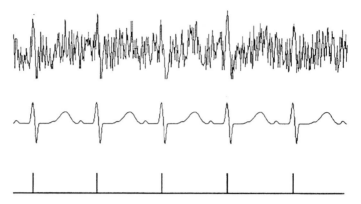

Figure 9.3 The top trace is the ECG of the center trace corrupted with random noise. The bottom trace provides fiducial marks that show the locations of the QRS peaks in the signal.

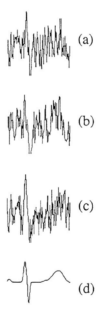

Figure 9.4 Summing the time-aligned signal epochs corrupted with random noise such as those in (a), (b), and (c), which were extracted from Figure 9.3, improves the signal-to-noise ratio. The result of averaging 100 of these ECG time epochs to improve the SNR by 10 is in (d).

9.2 SIGNAL AVERAGING AS A DIGITAL FILTER

Signal averaging is a kind of digital filtering process. The Fourier transform of the transfer function of an averager is composed of a series of discrete frequency components. Figure 9.5 shows how each of these components has the same spectral characteristics and amplitudes. Because of the appearance of its amplitude response, this type of filter is called a comb filter.

The width of each tooth decreases as the number of sweep repetitions increases. The desired signal has a frequency spectrum composed of discrete frequency components, a fundamental and harmonics. Noise, on the other hand, has a continuous distribution. As the bandwidth of each of the teeth of the comb decreases, this filter more selectively passes the fundamental and harmonics of the signal while rejecting the random noise frequencies that fall between the comb teeth. The signal averager, therefore, passes the signal while rejecting the noise.

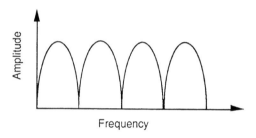

Figure 9.5 Fourier transform of a signal averager. As the number of sweeps increase, the width of each tooth of the comb decreases.

9.3 A TYPICAL AVERAGER

Figure 9.6 shows the block diagram of a typical averager. To average a signal such as the cortical response to an auditory stimulus, we stimulate the system (in this case, a human subject) with an auditory click to the stimulus input. Simultaneously, we provide a trigger derived from the stimulus that enables the summation of the sampled data (in this case, the EEG evoked by the stimulus) with the previous responses (time epochs or sweeps) stored in the buffer. When the averager receives the trigger pulse, it samples the EEG waveform at the selected rate, digitizes the signal, and sums the samples with the contents of a memory location corresponding to that sample interval (in the buffer). The process continues, stepping through the memory addresses until all addresses have been sampled. The sweep is terminated at this point. A new sweep begins with the next trigger and the cycle repeats until the desired number of sweeps have been averaged. The result of the averaging process is stored in the buffer which can then be displayed on a CRT as the averaged evoked response.

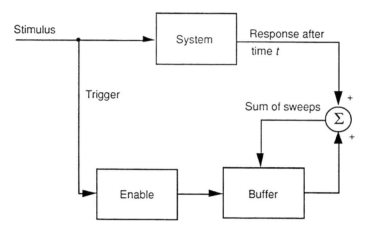

Figure 9.6 Block diagram of a typical signal averager.

9.4 SOFTWARE FOR SIGNAL AVERAGING

Figure 9.7 shows the flowchart of a program for averaging an ECG signal such as the one in Figure 9.3. The program uses a QRS detection algorithm to find a fiducial point at the peak of each QRS complex. Each time a QRS is detected, 128 new sample points are added to a buffer—64 points before and 64 points after the fiducial point.

9.5 LIMITATIONS OF SIGNAL AVERAGING

An important assumption made in signal averaging theory is that the noise is Gaussian. This assumption is not usually completely valid for biomedical signals. Also, if the noise distribution is related to the signal, misleading results can occur. If the fiducial point is derived from the signal itself, care must be taken to ensure that noise is not influencing the temporal location of the fiducial point. Otherwise, slight misalignment of each of the signal waveforms will lead to a low-pass filtering effect in the final result.

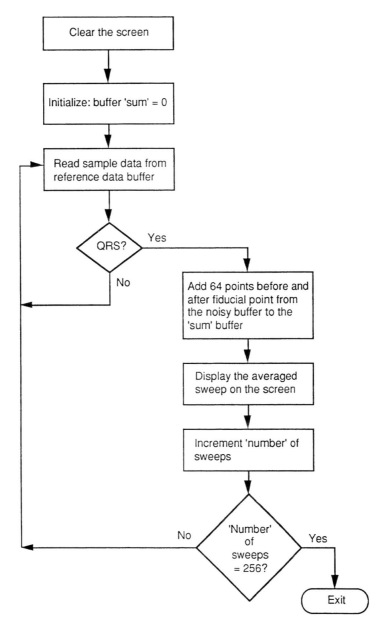

Figure 9.7 Flowchart of the signal averaging program.

9.6 LAB: ECG SIGNAL AVERAGING

Load UW DigiScope, select ad(V) Ops, then A(V)erage. This module is a demonstration of the use of signal averaging. The source file is always **average.dat** to which random, gaussian-distributed noise is added. A clean version of the data is preserved to use as a trigger signal. The averaged version of the data, displayed in the output channel, will build in size as successive waveforms are added. A summation of the accumulated signal traces is displayed, not the average. Thus you need to scale down the amplitude of the resultant signal in order to see the true average at the same amplitude scale factor as the original signal. Scaling of the output channel can be controlled as the traces are acquired. The down arrow key divides the amplitude by two each time it is struck. For example, while adding 16 heartbeats, you would need to strike the down arrow four times to divide the output by 16 so as to obtain the proper amplitude scale.

9.7 REFERENCES

Tompkins, W. J. and Webster, J. G. (eds.) 1981. *Design of Microcomputer-based Medical Instrumentation*. Englewood Cliffs, NJ: Prentice Hall.

9.8 STUDY QUESTIONS

9.1   Under what noise conditions will signal averaging fail to improve the SNR?
9.2   In a signal averaging application, the amplitude of uncorrelated noise is initially 16 times as large as the signal amplitude. How many sweeps must be averaged to give a resulting signal-to-noise ratio of 4:1?
9.3   After signal averaging 4096 EEG evoked responses, the signal-to-noise ratio is 4. Assuming that the EEG and noise sources are uncorrelated, what was the SNR before averaging?
9.4   In a signal averaging application, the noise amplitude is initially 4 times as large as the signal amplitude. How many sweeps must be averaged to give a resulting signal-to-noise ratio of 4:1?
9.5   In a signal averaging application, the signal caused by a stimulus and the noise are slightly correlated. The frequency spectra of the signal and noise overlap. Averaging 100 responses will improve the signal-to-noise ratio by what factor?

10

Data Reduction Techniques

Kok-Fung Lai

A typical computerized medical signal processing system acquires a large amount of data that is difficult to store and transmit. We need a way to reduce the data storage space while preserving the significant clinical content for signal reconstruction. In some applications, the process of reduction and reconstruction requires real-time performance (Jalaleddine et al., 1988).

A data reduction algorithm seeks to minimize the number of code bits stored by reducing the redundancy present in the original signal. We obtain the *reduction ratio* by dividing the number of bits of the original signal by the number saved in the compressed signal. We generally desire a high reduction ratio but caution against using this parameter as the sole basis of comparison among data reduction algorithms. Factors such as bandwidth, sampling frequency, and precision of the original data generally have considerable effect on the reduction ratio (Jalaleddine et al., 1990).

A data reduction algorithm must also represent the data with acceptable fidelity. In biomedical data reduction, we usually determine the clinical acceptability of the reconstructed signal through visual inspection. We may also measure the residual, that is, the difference between the reconstructed signal and the original signal. Such a numerical measure is the percent root-mean-square difference, PRD, given by

$$\text{PRD} = \left\{ \frac{\sum_{i=1}^{n} [x_{org}(i) - x_{rec}(i)]^2}{\sum_{i=1}^{n} [x_{org}(i)]^2} \right\}^{\frac{1}{2}} \times 100 \% \qquad (10.1)$$

where n is the number of samples and x_{org} and x_{rec} are samples of the original and reconstructed data sequences.

A *lossless* data reduction algorithm produces zero residual, and the reconstructed signal exactly replicates the original signal. However, clinically acceptable quality is neither guaranteed by a low nonzero residual nor ruled out by a high numerical residual (Moody et al., 1988). For example, a data reduction algorithm for an ECG recording may eliminate small-amplitude baseline drift. In this case, the residual contains negligible clinical information. The reconstructed ECG signal can thus be quite clinically acceptable despite a high residual.

In this chapter we discuss two classes of data reduction techniques for the ECG. The first class, significant-point-extraction, includes the turning point (TP) algorithm, AZTEC (Amplitude Zone Time Epoch Coding), and the Fan algorithm. These techniques generally retain samples that contain important information about the signal and discard the rest. Since they produce nonzero residuals, they are *lossy* algorithms. In the second class of techniques based on Huffman coding, variable-length code words are assigned to a given quantized data sequence according to frequency of occurrence. A predictive algorithm is normally used together with Huffman coding to further reduce data redundancy by examining a successive number of neighboring samples.

10.1 TURNING POINT ALGORITHM

The original motivation for the turning point (TP) algorithm was to reduce the sampling frequency of an ECG signal from 200 to 100 samples/s (Mueller, 1978). The algorithm developed from the observation that, except for QRS complexes with large amplitudes and slopes, a sampling rate of 100 samples/s is adequate.

TP is based on the concept that ECG signals are normally oversampled at four or five times faster than the highest frequency present. For example, an ECG used in monitoring may have a bandwidth of 50 Hz and be sampled at 200 sps in order to easily visualize the higher-frequency attributes of the QRS complex. Sampling theory tells us that we can sample such a signal at 100 sps. TP provides a way to reduce the effective sampling rate by half to 100 sps by selectively saving important signal points (i.e., the peaks and valleys or turning points).

The algorithm processes three data points at a time. It stores the first sample point and assigns it as the reference point X_0. The next two consecutive points become X_1 and X_2. The algorithm retains either X_1 or X_2, depending on which point preserves the turning point (i.e., slope change) of the original signal.

Figure 10.1(a) shows all the possible configurations of three consecutive sample points. In each frame, the solid point preserves the slope of the original three points. The algorithm saves this point and makes it the reference point X_0 for the next iteration. It then samples the next two points, assigns them to X_1 and X_2, and repeats the process.

We use a simple mathematical criterion to determine the saved point. First consider a *sign(x)* operation

$$sign(x) = \left\{ \begin{array}{ll} 0 & x = 0 \\ +1 & x > 0 \\ -1 & x < 0 \end{array} \right\} \qquad (10.2)$$

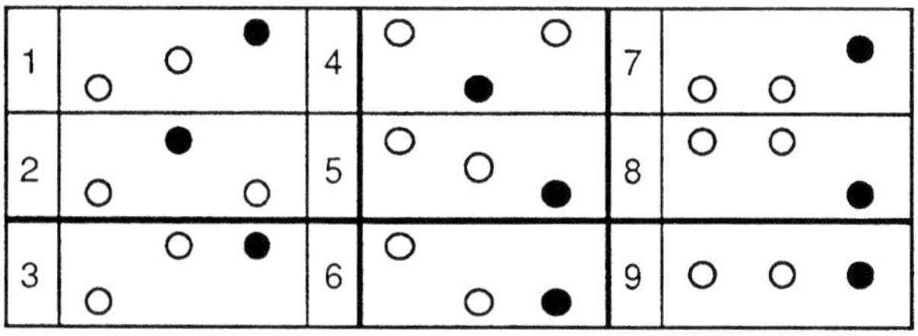

(a)

Pattern	$s_1 = sign(X_1 - X_0)$	$s_2 = sign(X_2 - X_1)$	NOT(s_1) OR ($s_1 + s_2$)	Saved sample
1	+1	+1	1	X_2
2	+1	−1	0	X_1
3	+1	0	1	X_2
4	−1	+1	0	X_1
5	−1	−1	1	X_2
6	−1	0	1	X_2
7	0	+1	1	X_2
8	0	−1	1	X_2
9	0	0	1	X_2

(b)

Figure 10.1 Turning point (TP) algorithm. (a) All possible 3-point configurations. Each frame includes the sequence of three points X_0, X_1, and X_2. The solid points are saved. (b) Mathematical criterion used to determine saved point.

We then obtain $s_1 = sign(X_1 - X_0)$ and $s_2 = sign(X_2 - X_1)$, where $(X_1 - X_0)$ and $(X_2 - X_1)$ are the slopes of the two pairs of consecutive points. If a slope is zero, this operator produces a zero result. For positive or negative slopes, it yields +1 or −1 respectively. A turning point occurs only when a slope changes from positive to negative or vice versa.

We use the logical Boolean operators, NOT and OR, as implemented in the C language to make the final judgment of when a turning point occurs. In the C language, NOT(c) = 1 if $c = 0$; otherwise NOT(c) = 0. Also logical OR means that (a OR b) = 0 only if a and b are both 0. Thus, we retain X_1 only if {NOT(s_1) OR ($s_1 + s_2$)} is zero, and save X_2 otherwise. In this expression, ($s_1 + s_2$) is the arithmetic sum of the signs produced by the *sign* function. The final effect of this processing is a Boolean decision whether to save X_1 or X_2. Point X_1 is saved only when the slope changes from positive to negative or vice versa. This computation could be easily done arithmetically, but the Boolean operation is computationally much faster.

Figure 10.2 shows the implementation of the TP algorithm in the C language. Figure 10.3 is an example of applying the TP algorithm to a synthesized ECG signal.

```
#define sign(x)  ( (x) ? ( (x > 0) ? 1 : -1 ) : 0 )
short *org, *tp ;          /* original and tp data */
short x0, x1, x2 ;         /* data points */
short s1, s2 ;                  /* signs */

x0 = *tp++ = *org++ ;      /* save the first sample */
while(there_is_sample) {

        x1 = *org++ ;
        x2 = *org++ ;
        s1 = sign(x1-x0) ;
        s2 = sign(x2-x1) ;
        *tp++ = x0 = ( !s1 || (s1+s2) ) ? x2 : x1 ;
}
```

Figure 10.2 C-language fragment showing TP algorithm implementation.

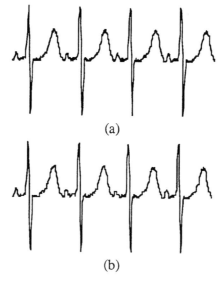

(a)

(b)

Figure 10.3 An example of the application of the TP algorithm. (a) Original waveform generated by the UW DigiScope **Genwave** function (see Appendix D). (b) Reconstructed signal after one application of the TP algorithm. Reduction ratio is 512:256, PRD = 7.78%.

The TP algorithm is simple and fast, producing a fixed reduction ratio of 2:1. After selectively discarding exactly half the sampled data, we can restore the original resolution by interpolating between pairs of saved data points.

A second application of the algorithm to the already reduced data increases the reduction ratio to 4:1. Using data acquired at a 200-sps rate, this produces compressed data with a 50-sps effective sampling rate. If the bandwidth of the acquired ECG is 50 Hz, this approach violates sampling theory since the effective sampling rate is less than twice the highest frequency present in the signal. The resulting reconstructed signal typically has a widened QRS complex and sharp edges that reduce its clinical acceptability. Another disadvantage of this algorithm is that the saved points do not represent equally spaced time intervals. This introduces short-term time distortion. However, this localized distortion is not visible when the reconstructed signal is viewed on the standard clinical monitors and paper recorders.

10.2 AZTEC ALGORITHM

Originally developed to preprocess ECGs for rhythm analysis, the AZTEC (Amplitude Zone Time Epoch Coding) data reduction algorithm decomposes raw ECG sample points into plateaus and slopes (Cox et al., 1968). It provides a sequence of line segments that form a piecewise-linear approximation to the ECG.

10.2.1 Data reduction

Figure 10.4 shows the complete flowchart for the AZTEC algorithm using C-language notation. The algorithm consists of two parts—line detection and line processing.

Figure 10.4(a) shows the line detection operation which makes use of zero-order interpolation (ZOI) to produce horizontal lines. Two variables V_{mx} and V_{mn} always reflect the highest and lowest elevations of the current line. Variable *LineLen* keeps track of the number of samples examined. We store a plateau if either the difference between V_{mxi} and V_{mni} is greater than a predetermined threshold V_{th} or if *LineLen* is greater than 50. The stored values are the length (*LineLen* − 1) and the average amplitude of the plateau $(V_{mx} + V_{mn})/2$.

Figure 10.4(b) shows the line processing algorithm which either produces a plateau or a slope depending on the value of the variable *LineMode*. We initialize *LineMode* to _PLATEAU in order to begin by producing a plateau. The production of an AZTEC slope begins when the number of samples needed to form a plateau is less than three. Setting *LineMode* to _SLOPE indicates that we have entered slope production mode. We then determine the direction or *sign* of the current slope by subtracting the previous line amplitude V_1 from the current amplitude V_{si}. We also reset the length of the slope T_{si}. The variable V_{si} records the current line amplitude so that any change in the direction of the slope can be tracked. Note that

V_{mxi} and V_{mni} are always updated to the latest sample before line detection begins. This forces ZOI to begin from the value of the latest sample.

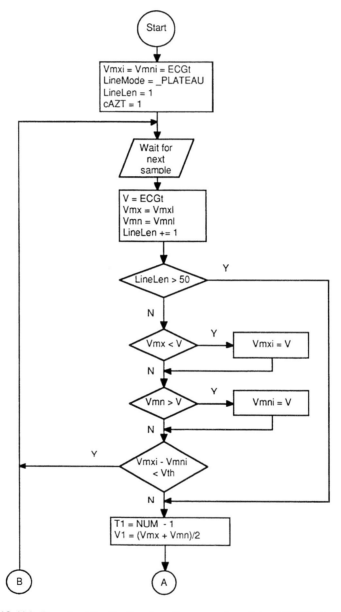

Figure 10.4(a) Flowchart for the line detection operation of the AZTEC algorithm.

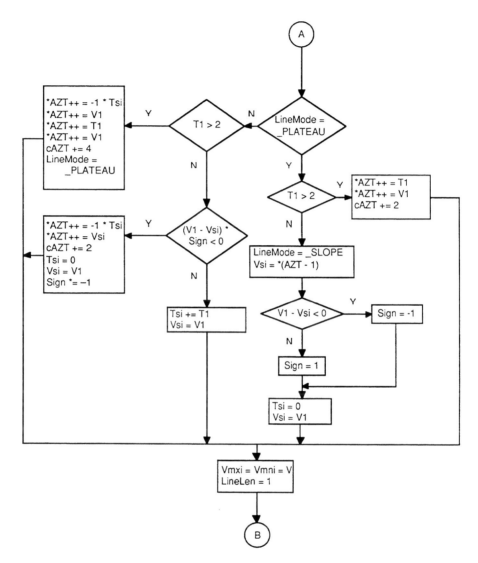

Figure 10.4(b) Flowchart of the line processing operation of the AZTEC algorithm.

When we reenter line processing with *LineMode* equal to _SLOPE, we either save or update the slope. The slope is saved either when a plateau of more than three samples can be formed or when a change in direction is detected. If we detect a new plateau of more than three samples, we store the current slope and the new plateau. For the slope, the stored values are its length T_{si} and its final elevation V_1.

Note that T_{si} is multiplied by -1 to differentiate a slope from a plateau (i.e., the minus sign serves as a flag to indicate a slope). We also store the length and the amplitude of the new plateau, then reset all parameters and return to plateau production.

If a change in direction is detected in the slope, we first save the parameters for the current slope and then reset $sign$, V_{si}, T_{si}, V_{mxi}, and V_{mni} to produce a new AZTEC slope. Now the algorithm returns to line detection but remains in slope production mode. When there is no new plateau or change of direction, we simply update the slope's parameters, T_{si} and V_{si}, and return to line detection with *LineMode* remaining set to _SLOPE.

AZTEC does not produce a constant data reduction ratio. The ratio is frequently as great as 10 or more, depending on the nature of the signal and the value of the empirically determined threshold.

10.2.2 Data reconstruction

The data array produced by the AZTEC algorithm is an alternating sequence of durations and amplitudes. A sample AZTEC-encoded data array is

$$\{18, 77, 4, 101, -5, -232, -4, 141, 21, 141\}$$

We reconstruct the AZTEC data by expanding the plateaus and slopes into discrete data points. For this particular example, the first two points represent a line 18 sample periods long at an amplitude of 77. The second set of two points represents another line segment 4 samples long at an amplitude of 101. The first value in the third set of two points is negative. Since this represents the length of a line segment, and we know that length must be positive, we recognize that this minus sign is the flag indicating that this particular set of points represents a line segment with nonzero slope. This line is five samples long beginning at the end of the previous line segment (i.e., amplitude of 101) and ending at an amplitude of -235. The next set of points is also a line with nonzero slope beginning at an amplitude of -235 and ending 4 sample periods later at an amplitude of 141.

This reconstruction process produces an ECG signal with steplike quantization, which is not clinically acceptable. The AZTEC-encoded signal needs postprocessing with a curve smoothing algorithm or a low-pass filter to remove its jagged appearance and produce more acceptable output.

The least square polynomial smoothing filter described in Chapter 5 is an easy and fast method for smoothing the signal. This family of filters fits a parabola to an odd number $(2L + 1)$ of input data points. Taking $L = 3$, we obtain

$$p_k = \frac{1}{21}\left(-2x_{k-3} + 3x_{k-2} + 6x_{k-1} + 7x_k + 6x_{k+1} + 3x_{k+2} - 2x_{k+3}\right) \quad (10.3)$$

where p_k is the new data point and x_k is the expanded AZTEC data. The smoothing function acts as a low-pass filter to reduce the discontinuities. Although this produces more acceptable output, it also introduces amplitude distortion.

Figure 10.5 shows examples of the AZTEC algorithm applied to an ECG.

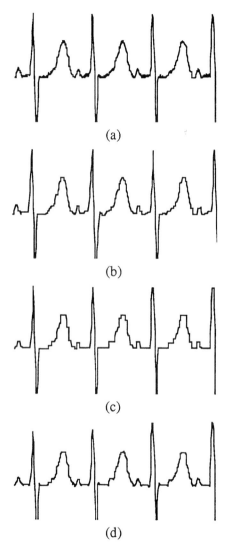

(a)

(b)

(c)

(d)

Figure 10.5 Examples of AZTEC applications. (a) Original waveform generated by the UW DigiScope **Genwave** function (see Appendix D). (b) Small threshold, reduction ratio = 512:233, PRD = 24.4%. (c) Large threshold, reduction ratio = 512:153, PRD = 28.1%. (d) Smoothed signal from (c), $L = 3$, PRD = 26.3%.

10.2.3 CORTES algorithm

The CORTES (Coordinate Reduction Time Encoding System) algorithm is a hybrid of the TP and AZTEC algorithms (Abenstein and Tompkins, 1982; Tompkins and Webster, 1981). It attempts to exploit the strengths of each while sidestepping the weaknesses. CORTES uses AZTEC to discard clinically insignificant data in the isoelectric region with a high reduction ratio and applies the TP algorithm to the clinically significant high-frequency regions (QRS complexes). It executes the AZTEC and TP algorithms in parallel on the incoming ECG data.

Whenever an AZTEC line is produced, the CORTES algorithm decides, based on the length of the line, whether the AZTEC data or the TP data are to be saved. If the line is longer than an empirically determined threshold, it saves the AZTEC line. Otherwise it saves the TP data points. Since TP is used to encode the QRS complexes, only AZTEC plateaus, not slopes, are implemented.

The CORTES algorithm reconstructs the signal by expanding the AZTEC plateaus and interpolating between each pair of the TP data points. It then applies parabolic smoothing to the AZTEC portions to reduce discontinuities.

10.3 FAN ALGORITHM

Originally used for ECG telemetry, the Fan algorithm draws lines between pairs of starting and ending points so that all intermediate samples are within some specified error tolerance, ε (Bohs and Barr, 1988). Figure 10.6 illustrates the principles of the Fan algorithm. We start by accepting the first sample X_0 as the nonredundant permanent point. It functions as the origin and is also called the originating point. We then take the second sample X_1 and draw two slopes $\{U_1, L_1\}$. U_1 passes through the point $(X_0, X_1 + \varepsilon)$, and L_1 passes through the point $(X_0, X_1 - \varepsilon)$. If the third sample X_2 falls within the area bounded by the two slopes, we generate two new slopes $\{U_2, L_2\}$ that pass through points $(X_0, X_2 + \varepsilon)$ and $(X_0, X_2 - \varepsilon)$. We compare the two pairs of slopes and retain the most converging (restrictive) slopes (i.e., $\{U_1, L_2\}$ in our example). Next we assign the value of X_2 to X_1 and read the next sample into X_2. As a result, X_2 always holds the most recent sample and X_1 holds the sample immediately preceding X_2. We repeat the process by comparing X_2 to the values of the most convergent slopes. If it falls outside this area, we save the length of the line T and its final amplitude X_1 which then becomes the new originating point X_0, and the process begins anew. The sketch of the slopes drawn from the originating sample to future samples forms a set of radial lines similar to a fan, giving this algorithm its name.

When adapting the Fan algorithm to C-language implementation, we create the variables, X_{U1}, X_{L1}, X_{U2}, and X_{L2}, to determine the bounds of X_2. From Figure 10.6(b), we can show that

$$X_{U2} = \frac{X_{U1} - X_0}{T} + X_{U1} \tag{10.4a}$$

and

$$X_{L2} = \frac{X_{L1} - X_0}{T} + X_{L1} \qquad (10.4b)$$

where $T = t_T - t_0$.

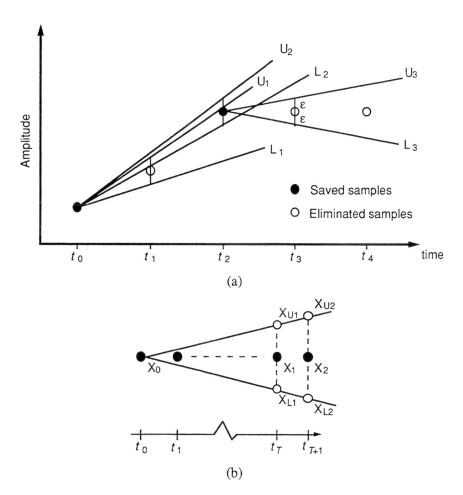

(a)

(b)

Figure 10.6 Illustration of the Fan algorithm. (a) Upper and lower slopes (U and L) are drawn within error threshold ε around sample points taken at $t_1, t_2, \dots$ (b) Extrapolation of X_{U2} and X_{L2} from X_{U1}, X_{L1}, and X_0.

Figure 10.7 shows the C-language fragment that implements the Fan algorithm. Figure 10.8 shows an example of the Fan algorithm applied to an ECG signal.

```
short X0, X1, X2 ;                /* sample points */
short XU2, XL2, XU1, XL1 ;        /* variable to determine bounds */
short Epsilon ;                   /* threshold */
short *org, *fan ;                /* original and Fan data */
short T ;                         /* length of line */
short V2 ;                          /* sample point */

/* initialize all variables */
X0 = *org++ ;                     /* the originating point */
X1 = *org++ ;                     /* the next point */
T = 1 ;                           /* line length is initialize to 1 */
XU1 = X1 + Epsilon ;              /* upper bound of X1 */
XL1 = X1 - Epsilon ;              /* lower bound of X1 */
*fan++ = X0 ;                     /* save the first permanent point */

while( there_is_data )    {

        V2 = *org++ ;                      /* get next sample point */
        XU2 = (XU1 - X0)/T + XU1;          /* upper bound of X2 */
        XL2 = (XL2 - X0)/T + XL2;  /* lower bound of X2 */

        if( X2 <= XU2 && X2 >= XL2 )     {        /* within bound */

                /* obtain the most restrictive bound */
                XU2 = (XU2 < X2 + Epsilon) ? XU2 : X2 + Epsilon ;
                XL2 = (XL2 > X2 - Epsilon) ? XL2 : X2 - Epsilon ;

                T++ ;  /* increment line length */
                X1 = X2 ;       /* X1 hold sample preceding X2 */
        }

        else    {       /* X2 out of bound, save line */

                *fan++ = T ;  /* save line length */
                *fan++ = X1 ;           /* save final amplitude */

                /* reset all variables */
                X0 = X1 ;
                X1 = X2 ;
                T = 1 ;
                XU1 = X1 + Epsilon ;
                XL1 = X1 - Epsilon ;
        }
}
```

Figure 10.7 Fragment of C-language program for implementation of the Fan algorithm.

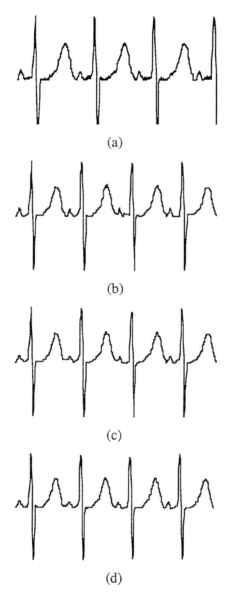

(a)

(b)

(c)

(d)

Figure 10.8 Examples of Fan algorithm applications. (a) Original waveform generated by the UW DigiScope Genwave function (see Appendix D). (b) Small tolerance, reduction ratio = 512:201 PRD = 5.6%. (c) Large tolerance, reduction ratio = 512:155, PRD = 7.2%. (d) Smoothed signal from (c), $L = 3$, PRD = 8.5%.

We reconstruct the compressed data by expanding the lines into discrete points. The Fan algorithm guarantees that the error between the line joining any two permanent sample points and any actual (redundant) sample along the line is less than or equal to the magnitude of the preset error tolerance. The algorithm's reduction ratio depends on the error tolerance. When compared to the TP and AZTEC algorithms, the Fan algorithm produces better signal fidelity for the same reduction ratio (Jalaleddine et al., 1990).

Three algorithms based on Scan-Along Approximation (SAPA) techniques (Ishijima et al., 1983; Tai, 1991) closely resemble the Fan algorithm. The SAPA-2 algorithm produces the best results among all three algorithms. As in the Fan algorithm, SAPA-2 guarantees that the deviation between the straight lines (reconstructed signal) and the original signal never exceeds a preset error tolerance.

In addition to the two slopes calculated in the Fan algorithm, SAPA-2 calculates a third slope called the center slope between the originating sample point and the actual future sample point. Whenever the center slope value does not fall within the boundary of the two converging slopes, the immediate preceding sample is taken as the originating point. Therefore, the only apparent difference between SAPA-2 and the Fan algorithm is that the SAPA-2 uses the center slope criterion instead of the actual sample value criterion.

10.4 HUFFMAN CODING

Huffman coding exploits the fact that discrete amplitudes of quantized signal do not occur with equal probability (Huffman, 1952). It assigns variable-length code words to a given quantized data sequence according to their frequency of occurrence. Data that occur frequently are assigned shorter code words.

10.4.1 Static Huffman coding

Figure 10.9 illustrates the principles of Huffman coding. As an example, assume that we wish to transmit the set of 28 data points

$$\{1, 1, 1, 1, 1, 1, 1, 2, 2, 2, 2, 2, 2, 3, 3, 3, 3, 3, 4, 4, 4, 4, 5, 5, 5, 6, 6, 7\}$$

The set consists of seven distinct quantized levels, or *symbols*. For each symbol, S_i, we calculate its probability of occurrence P_i by dividing its frequency of occurrence by 28, the total number of data points. Consequently, the construction of a Huffman code for this set begins with seven nodes, one associated with each P_i. At each step we sort the P_i list in descending order, breaking the ties arbitrarily. The two nodes with smallest probability, P_i and P_j, are merged into a new node with probability $P_i + P_j$. This process continues until the probability list contains a single value, 1.0, as shown in Figure 10.9(a).

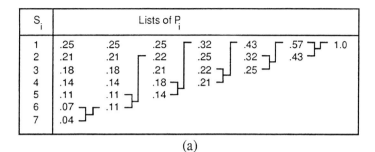

(a)

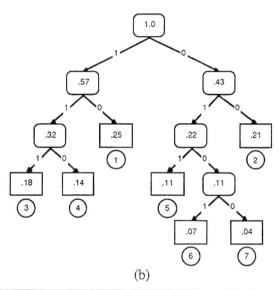

(b)

Symbols, S_i	3-bit binary code	Probability of occurrence, P_i	Huffman code
1	001	0.25	10
2	010	0.21	00
3	011	0.18	111
4	100	0.14	110
5	101	0.11	011
6	110	0.07	0101
7	111	0.04	0100

(c)

Figure 10.9 Illustration of Huffman coding. (a) At each step, P_i are sorted in descending order and the two lowest P_i are merged. (b) Merging operation depicted in a binary tree. (c) Summary of Huffman coding for the data set.

The process of merging nodes produces a binary tree as in Figure 10.9(b). When we merge two nodes with probability $P_i + P_j$, we create a parent node with two children represented by P_i and P_j. The root of the tree has probability 1.0. We obtain the Huffman code of the symbols by traversing down the tree, assigning 1 to the left child and 0 to the right child. The resulting code words have the *prefix property* (i.e., no code word is a proper prefix of any other code word). This property ensures that a coded message is uniquely decodable without the need for lookahead. Figure 10.9(c) summarizes the results and shows the Huffman codes for the seven symbols. We enter these code word mappings into a translation table and use the table to pad the appropriate code word into the output bit stream in the reduction process.

The reduction ratio of Huffman coding depends on the distribution of the source symbols. In our example, the original data requires three bits to represent the seven quantized levels. After Huffman coding, we can calculate the expected code word length

$$E[l] = \sum_{i=1}^{7} l_i P_i \qquad (10.5)$$

where l_i represents the length of Huffman code for the symbols. This value is 2.65 in our example, resulting in an expected reduction ratio of 3:2.65.

The reconstruction process begins at the root of the tree. If bit 1 is received, we traverse down the left branch, otherwise the right branch. We continue traversing until we reach a node with no child. We then output the symbol corresponding to this node and begin traversal from the root again.

The reconstruction process of Huffman coding perfectly recovers the original data. Therefore it is a lossless algorithm. However, a transmission error of a single bit may result in more than one decoding error. This propagation of transmission error is a consequence of all algorithms that produce variable-length code words.

10.4.2 Modified Huffman coding

The implementation of Huffman coding requires a translation table, where each source symbol is mapped to a unique code word. If the original data were quantized into 16-bit numbers, the table would need to contain 2^{16} records. A table of this size creates memory problems and processing inefficiency.

In order to reduce the size of the translation table, the modified Huffman coding scheme partitions the source symbols into a frequent set and an infrequent set. For all the symbols in the frequent set, we form a Huffman code as in the static scheme. We then use a special code word as a prefix to indicate any symbol from the infrequent set and attach a suffix corresponding to the ordinary binary encoding of the symbol.

Assume that we are given a data set similar to the one before. Assume also that we anticipate quantized level 0 to appear in some future transmissions. We may decide to partition the quantized levels {0, 7} into the infrequent set. We then apply Huffman coding as before and obtain the results in Figure 10.10. Note that quantized levels in the infrequent set have codes with prefix 0100, making their code length much longer than those of the frequent set. It is therefore important to keep the probability of the infrequent set sufficiently small to achieve a reasonable reduction ratio.

Some modified Huffman coding schemes group quantized levels centered about 0 into the frequent set and derive two prefix codes for symbols in the infrequent set. One prefix code denotes large positive values and the other denotes large negative values.

10.4.3 Adaptive coding

Huffman coding requires a translation table for encoding and decoding. It is necessary to examine the entire data set or portions of it to determine the data statistics. The translation table must also be transmitted or stored for correct decoding.

An adaptive coding scheme attempts to build the translation table as data are presented. A dynamically derived translation table is sensitive to the variation in local statistical information. It can therefore alter its code words according to local statistics to maximize the reduction ratio. It also achieves extra space saving because there is no need for a static table.

An example of an adaptive scheme is the Lempel-Ziv-Welch (LZW) algorithm. The LZW algorithm uses a fixed-size table. It initializes some positions of the table for some chosen data sets. When it encounters new data, it uses the uninitialized positions so that each unique data word is assigned its own position. When the table is full, the LZW algorithm reinitializes the oldest or least-used position according to the new data. During data reconstruction, it incrementally rebuilds the translation table from the encoded data.

Symbols, S_i	3-bit binary code	Probability of occurrence, P_i	Huffman code
0	000	0.00	0100000
1	001	0.25	10
2	010	0.21	00
3	011	0.18	111
4	100	0.14	110
5	101	0.11	011
6	110	0.07	0101
7	111	0.04	0100111

Figure 10.10 Results of modified Huffman coding. Quantized levels {0, 7} are grouped into the infrequent set.

10.4.4 Residual differencing

Typically, neighboring signal amplitudes are not statistically independent. Conceptually we can decompose a sample value into a part that is correlated with past samples and a part that is uncorrelated. Since the intersample correlation corresponds to a value predicted using past samples, it is redundant and removable. We are then left with the uncorrelated part which represents the prediction error or residual signal. Since the amplitude range of the residual signal is smaller than that of the original signal, it requires less bits for representation. We can further reduce the data by applying Huffman coding to the residual signal. We briefly describe two ECG reduction algorithms that make use of residual differencing.

Ruttimann and Pipberger (1979) applied modified Huffman coding to residuals obtained from prediction and interpolation. In prediction, sample values are obtained by taking a linearly weighted sum of an appropriate number of past samples

$$x'(nT) = \sum_{k=1}^{p} a_k\, x(nT - kT) \tag{10.6}$$

where $x(nT)$ are the original data, $x'(nT)$ are the predicted samples, and p is the number of samples employed in prediction. The parameters a_k are chosen to minimize the expected mean squared error $E[(x - x')^2]$. When $p = 1$, we choose $a_1 = 1$ and say that we are taking the *first difference* of the signal. Preliminary investigations on test ECG data showed that there was no substantial improvement by using predictors higher than second order (Ruttimann et al., 1976). In interpolation, the estimator of the sample value consists of a linear combination of past and future samples. The results for the predictor indicated a second-order estimator to be sufficient. Therefore, the interpolator uses only one past and one future sample

$$x'(n) = ax(nT - T) + bx(nT + T) \tag{10.7}$$

where the coefficients a and b are determined by minimizing the expected mean squared error. The residuals of prediction and interpolation are encoded using a modified Huffman coding scheme, where the frequent set consists of some quantized levels centered about zero. Encoding using residuals from interpolation resulted in higher reduction ratio of approximately 7.8:1.

Hamilton and Tompkins (1991a, 1991b) exploited the fact that a typical ECG signal is composed of a repeating pattern of beats with little change from beat to beat. The algorithm calculates and updates an average beat estimate as data are presented. When it detects a beat, it aligns and subtracts the detected beat from the average beat. The residual signal is Huffman coded and stored along with a record of the beat locations. Finally, the algorithm uses the detected beat to update the average beat estimate. In this scheme, the estimation of beat location and quantizer location can significantly affect reduction performance.

10.4.5 Run-length encoding

Used extensively in the facsimile technology, run-length encoding exploits the high degree of correlation that occurs in successive bits in the facsimile bit stream. A bit in the facsimile output may either be 1 or 0, depending on whether it is a black or white pixel. On a typical document, there are clusters of black and white pixels that give rise to this high correlation. Run-length encoding simply transforms the original bit stream into the string $\{v_1, l_1, v_2, l_2, ...\}$ where v_i are the values and l_i are the lengths. The observant reader will quickly recognize that both AZTEC and the Fan algorithm are special cases of run-length encoding.

Take for example the output stream $\{1, 1, 1, 1, 1, 3, 3, 3, 3, 0, 0, 0\}$ with 12 elements. The output of run-length encoding $\{1, 5, 3, 4, 0, 3\}$ contains only six elements. Further data reduction is possible by applying Huffman coding to the output of run-length encoding.

10.5 LAB: ECG DATA REDUCTION ALGORITHMS

This lab explores the data reduction techniques reviewed in this chapter. Load UW DigiScope according to the directions in Appendix D.

10.5.1 Turning point algorithm

From the ad(V) Ops menu, select c(O)mpress and then (T)urn pt. The program compresses the waveform displayed on the top channel using the TP algorithm, then decompresses, reconstructs using interpolation, and displays the results on the bottom channel. Perform the TP algorithm on two different ECGs read from files and on a sine wave and a square wave. Observe

1. Quality of the reconstructed signal
2. Reduction ratio
3. Percent root-mean-square difference (PRD)
4. Power spectra of original and reconstructed signals.

Tabulate and summarize all your observations.

10.5.2 AZTEC algorithm

Repeat section 10.5.1 for the AZTEC algorithm by selecting (A)ztec from the COMPRESS menu. Using at least three different threshold values (try 1%, 5%, and 15% of the full-scale peak-to-peak value), observe and comment on the items in the list in section 10.5.1. In addition, summarize the quality of the reconstructed signals both before and after applying the smoothing filter. Tabulate and summarize all your observations.

10.5.3 Fan algorithm

Repeat section 10.5.2 for the Fan algorithm by selecting (F)an from the COMPRESS menu. What can you deduce from comparing the performance of the Fan algorithm with that of the AZTEC algorithm? Tabulate and summarize all your observations.

10.5.4 Huffman coding

Select (H)uffman from the COMPRESS menu. Select (R)un in order to Huffman encode the signal that is displayed on the top channel. Do not use first differencing at this point in the experiment. Record the reduction ratio. Note that this reduction ratio does not include the space needed for the translation table which must be stored or transmitted. What can you deduce from the PRD? Select (W)rite table to write the Huffman data into a file. You may view the translation table later with the DOS type command after exiting from SCOPE.

Load a new ECG waveform and repeat the steps above. When you select (R)un, the program uses the translation table derived previously to code the signal. What can you deduce from the reduction ratio? After deriving a new translation table using (M)ake from the menu, select (R)un again and comment on the new reduction ratio.

Select (M)ake again and use first differencing to derive a new Huffman code. Is there a change in the reduction ratio using this newly derived code? Select (W)rite table to write the Huffman data into a file. Now reload the first ECG waveform that you used. Without deriving a new Huffman code, observe the reduction ratio obtained. Comment on your observations.

Exit from the SCOPE program to look at the translation tables that you generated. What comments can you make regarding the overhead involved in storing a translation table?

10.6 REFERENCES

Abenstein, J. P. and Tompkins, W. J. 1982. New data-reduction algorithm for real-time ECG analysis, *IEEE Trans. Biomed. Eng.*, **BME-29**: 43–48.

Bohs, L. N. and Barr, R. C. 1988. Prototype for real-time adaptive sampling using the Fan algorithm, *Med. & Biol. Eng. & Comput.*, **26**: 574–83.

Cox, J. R., Nolle, F. M., Fozzard, H. A., and Oliver, G. C. Jr. 1968. AZTEC: a preprocessing program for real-time ECG rhythm analysis. *IEEE Trans. Biomed. Eng.*, **BME-15**: 128–29.

Hamilton, P. S., and Tompkins, W. J. 1991a. Compression of the ambulatory ECG by average beat subtraction and residual differencing. *IEEE Trans. Biomed. Eng.*, **BME-38**(3): 253–59.

Hamilton, P. S., and Tompkins, W. J. 1991b. Theoretical and experimental rate distortion performance in compression of ambulatory ECGs. *IEEE Trans. Biomed. Eng.*, **BME-38**(3): 260–66.

Huffman, D. A. 1952. A method for construction of minimum-redundancy codes. *Proc. IRE*, **40**: 1098–1101.

Ishijima, M., Shin, S. B., Hostetter, G. H., and Skalansky, J. 1983. Scan-along polygonal approximation for data compression of electrocardiograms, *IEEE Trans. Biomed. Eng.*, **BME-30**: 723–29.

Jalaleddine, S. M. S., Hutchens, C. G., Coberly, W. A., and Strattan, R. D. 1988. Compression of Holter ECG data. *Biomedical Sciences Instrumentation*, **24**: 35–45.

Jalaleddine, S. M. S., Hutchens. C. G., and Strattan, R. D. 1990. ECG data compression techniques — A unified approach. *IEEE Trans. Biomed. Eng.*, **BME-37**: 329–43.

Moody, G. B., Soroushian., and Mark, R. G. 1988. ECG data compression for tapeless ambulatory monitors. *Computers in Cardiology*, 467–70.

Mueller, W. C. 1978. Arrhythmia detection program for an ambulatory ECG monitor. *Biomed. Sci. Instrument.*, **14**: 81–85.

Ruttimann, U. E. and Pipberger, H. V. 1979. Compression of ECG by prediction or interpolation and entropy encoding. *IEEE Trans. Biomed. Eng.*, **BME-26**: 613–23.

Ruttimann, U. E., Berson, A. S., and Pipberger, H. V. 1976. ECG data compression by linear prediction. *Proc. Comput. Cardiol.*, 313–15.

Tai, S. C. 1991. SLOPE — a real-time ECG data compressor. *Med. & Biol. Eng. & Comput.*, 175–79.

Tompkins, W. J. and Webster, J. G. (eds.) 1981. *Design of Microcomputer-based Medical Instrumentation*. Englewood Cliffs, NJ: Prentice Hall.

10.7 STUDY QUESTIONS

10.1   Explain the meaning of lossless and lossy data compression. Classify the four data reduction algorithms described in this chapter into these two categories.

10.2   Given the following data: {15, 10, 6, 7, 5, 3, 4, 7, 15, 3}, produce the data points that are stored using the TP algorithm.

10.3   Explain why an AZTEC reconstructed waveform is unacceptable to a cardiologist. Suggest ways to alleviate the problem.

10.4   The Fan algorithm can be applied to other types of biomedical signals. List the desirable characteristics of the biomedical signal that will produce satisfactory results using this algorithm. Give an example of such a signal.

10.5   Given the following data set: {a, a, a, a, b, b, b, b, b, c, c, c, d, d, e}, derive the code words for the data using Huffman coding. What is the average code word length?

10.6   Describe the advantages and disadvantages of modified Huffman coding.

10.7   Explain why it is desirable to apply Huffman coding to the residuals obtained by subtracting the estimated sample points from the original sample points.

10.8   Data reduction can be performed using parameter extraction techniques. A particular characteristic or parameter of the signal is extracted and transmitted in place of the original signal. Draw a block diagram showing the possible configuration for such a system. Your block diagram should include the compression and the reconstruction portions. What are the factors governing the success of these techniques?

10.9   Does the TP algorithm (a) produce significant time-base distortion over a very long time, (b) save every turning point (i.e., peak or valley) in a signal, (c) provide data reduction of 4-to-1 if applied twice to a signal without violating sampling theory, (d) provide for exactly reconstructing the original signal, (e) perform as well as AZTEC for electroencephalography (EEG)? Explain your answers.

10.10 Which of the following are characteristic of a Huffman coding algorithm? (a) Guarantees more data reduction on an ECG than AZTEC; (b) Cannot perfectly reconstruct the sampled data points (within some designated error range); (c) Is a variable-length code; (d) Is derived directly from Morse code; (e) Uses ASCII codes for the most frequent A/D values; (f) Requires advance knowledge of the frequency of occurrence of data patterns; (g) Includes as part of the algorithm self-correcting error checks.

10.11   After application of the TP algorithm, what data sequence would be saved if the data sampled by an analog-to-digital converter were: (a) {20, 40, 20, 40, 20, 40, 20, 40}, (b) {50, 40, 50, 20, 30, 40}, (c) {50, 50, 40, 30, 40, 50, 40, 30, 40, 50, 50, 40}, (d) {50, 25, 50, 25, 50, 25, 50, 25}?

10.12   After application of the TP algorithm on a signal, the data points saved are {50, 70, 30, 40}. If you were to reconstruct the original data set, what is the data sequence that would best approximate it?

10.13   The graph below shows a set of 20 data points sampled from an analog-to-digital converter. At the top of the chart are the numerical values of the samples. The solid lines represent AZTEC encoding of this sampled signal.

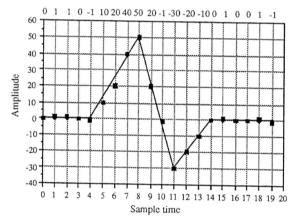

(a) List the data array that represents the AZTEC encoding of this signal.

(b) How much data reduction does AZTEC achieve for this signal?

(c) Which data points in the following list of raw data that would be saved if the Turning Point algorithm were applied to this signal?

$$0\ 1\ 1\ 0\ -1\ 10\ 20\ 40\ 50\ 20\ -1\ -30\ -20\ -10\ 0\ 1\ 0\ 0\ 1\ -1$$

(d) If this signal were encoded with a Huffman-type variable-bit-length code with the following four bit patterns as part of the set of codes, indicate which amplitude value you would assign to each pattern.

Code	Amplitude value
1	
01	
001	
0001	

(e) How much data reduction does each algorithm provide (assuming that no coding table needs to be stored for Huffman coding)?

10.14   AZTEC encodes a signal as {2, 50, –4, 30, –4, 50, –4, 30, –4, 50, 2, 50}. How many data points were originally sampled?

10.15   After applying the AZTEC algorithm to a signal, the saved data array is {2, 0, –3, 80, –3, –30, –3, 0, 3, 0}. Draw the waveform that AZTEC would reconstruct from these data.

10.16 AZTEC encodes a signal from an 8-bit analog-to-digital converter as {2, 50, –4, 30, –6, 50, –6, 30, –4, 50, 2, 50}. (a) What is the amount of data reduction? (b) What is the peak-to-peak amplitude of a signal reconstructed from these data?

10.17 AZTEC encodes a signal from an 8-bit analog-to-digital converter as {3, 100, –5, 150, –5, 50, 5, 100, 2, 100}. The TP algorithm is applied to the same original signal. How much more data reduction does AZTEC achieve on the same signal compared to TP?

10.18 The graph below shows a set of 20 data points of an ECG sampled with an 8-bit analog-to-digital converter.

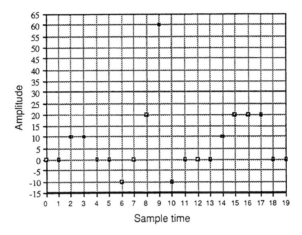

(a) Draw a Huffman binary tree similar to the one in Figure 10.9(b) including the probabilities of occurrence for this set of data.

(b) (5 points) From the binary tree, assign appropriate Huffman codes to the numbers in the data array:

Number	Huffman code
–10	
0	
10	
20	
60	

(c) Assuming that the Huffman table does not need to be stored, how much data reduction is achieved with Huffman coding of this sampled data set? (Note: Only an integral number of bytes may be stored.)

(d) Decode the following Huffman-coded data and list the sample points that it represents:

0 1 0 1 0 1 1 0 0 0 1

11

Other Time- and Frequency-Domain Techniques

Dorin Panescu

A biomedical signal is often corrupted by noise (e.g., powerline interference, muscle or motion artifacts, RF interference from electrosurgery or diathermy apparatus). Therefore, it is useful to know the frequency spectrum of the corrupting signal in order to be able to design a filter to eliminate it. If we want to find out, for example, how well the patient's cardiac output is correlated with the area of the QRS complex, then we need to use proper correlation techniques. This chapter presents time and frequency-domain techniques that might be useful for situations such as those exemplified above.

11.1 THE FOURIER TRANSFORM

The digital computer algorithm for Fourier analysis called the fast Fourier transform (FFT) serves as a basic tool for frequency-domain analysis of signals.

11.1.1 The Fourier transform of a discrete nonperiodic signal

Assuming that a discrete-time aperiodic signal exists as a sequence of data sampled from an analog prototype with a sampling period of T, the angular sampling frequency being $\omega_S = 2\pi/T$, we can write this signal in the time domain as a series of weighted Dirac functions. Thus

$$x(t) = \sum_{n=-\infty}^{\infty} x(n)\, \delta(t - nT) \tag{11.1}$$

216

The Fourier transform of this expression is

$$X(\omega) = \int_{-\infty}^{\infty} x(t)\, e^{-j\omega t} dt \tag{11.2}$$

or

$$X(\omega) = \int_{-\infty}^{\infty} \sum_{n=-\infty}^{\infty} x(n)\ \delta(t - nT)\, e^{-j\omega t} dt \tag{11.3a}$$

The ordering of integration and summation can be changed to give

$$X(\omega) = \sum_{n=-\infty}^{\infty} x(n) \int_{-\infty}^{\infty} \delta(t - nT) e^{-j\omega t}\, dt \tag{11.3b}$$

Thus we obtain

$$X(\omega) = \sum_{n=-\infty}^{\infty} x(n)\, e^{-j\omega n T} \tag{11.3c}$$

And similarly, we can find that the inverse Fourier transform is

$$x(n) = \frac{T}{2\pi} \int_{0}^{\omega_s} X(\omega)\, e^{j\omega n T}\, d\omega \tag{11.4}$$

One of the important properties of the Fourier transform, which is shown in Figure 11.1(b), is its repetition at intervals of the sampling frequency in both positive and negative directions. Also it is remarkable that the components in the interval $0 < \omega < \omega_s/2$ are the complex conjugates of the components in the interval $\omega_s/2 < \omega < \omega_s$. It is modern practice to use normalized frequencies, which means that the sampling period T is taken to be 1. Therefore, the Fourier transform pair for discrete signals, considering normalized frequencies, is

$$X(\omega) = \sum_{n=-\infty}^{\infty} x(n)\, e^{-j\omega n} \tag{11.5a}$$

$$x(n) = \frac{1}{2\pi} \int_{-\pi}^{\pi} X(\omega)\, e^{j\omega n}\, d\omega \tag{11.5b}$$

We observe that this kind of Fourier transform is continuous, and it repeats at intervals of the sampling frequency.

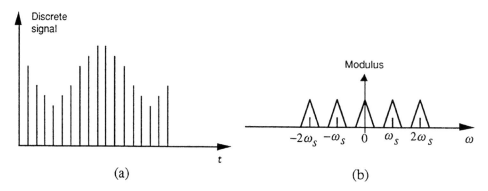

Figure 11.1 (a) A discrete-time signal, and (b) the modulus of its Fourier transform. Symmetry about $\omega_S/2$, due to the sampling process, is illustrated.

11.1.2 The discrete Fourier transform for a periodic signal

The discrete Fourier transform (DFT) is the name given to the calculation of the Fourier series coefficients for a discrete periodic signal. The operations are similar to the calculation of Fourier coefficients for a periodic signal, but there are also certain marked differences. The first is that the integrals become summations in the discrete time domain. The second difference is that the transform evaluates only a finite number of complex coefficients, the total being equal to the original number of data points in one period of original signal. Because of this, each spectral line is regarded as the k-*th* harmonic of the basic period in the data rather than identifying with a particular frequency expressed in Hz or radian/s. Algebraically, the forward and reverse transforms are expressed as

$$X(k) = \sum_{n=0}^{N-1} x(n)\, e^{\frac{-jkn\,2\pi}{N}} \qquad (11.6a)$$

$$x(n) = \frac{1}{N} \sum_{k=0}^{N-1} X(k)\, e^{\frac{jkn\,2\pi}{N}} \qquad (11.6b)$$

Figure 11.2 shows a discrete periodic signal and the real and imaginary parts of its DFT. The first spectral line ($k = 0$) gives the amplitude of the dc component in the signal, and the second line corresponds to that frequency which represents one cycle in N data points. This frequency is $2\pi/N$. The N-th line corresponds to the sampling frequency of the discrete N-sample data sequence per period and the $k = N/2$-th line corresponds to the Nyquist frequency. Using the symmetry of the DFT, algorithms for fast computation have been developed. Also, the symmetry has two important implications. The first is that the transformation will yield N unique complex spectral lines. The second is that half of these are effectively redundant because all of the information contained in a real time domain signal is contained within the first $N/2$ complex spectral lines. These facts permitted the development of the Fast Fourier Transform (FFT), which is presented in the next section.

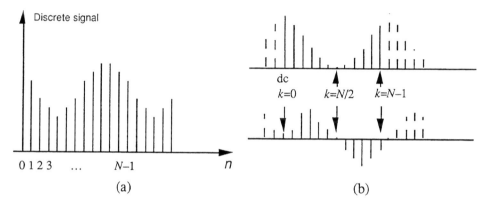

Figure 11.2 (a) A discrete periodic signal and (b) the real and imaginary parts of its DFT.

11.1.3 The fast Fourier transform

For a complete discussion of this subject, see Oppenheim and Schafer (1975). The term FFT applies to any computational algorithm by which the discrete Fourier transform can be evaluated for a signal consisting of N equally spaced samples, with N usually being a power of two. To increase the computation efficiency, we must divide the DFT into successively smaller DFTs. In this process we will use the symmetry and the periodicity properties of the complex exponential

$$W_N^{kn} = e^{-j(2\pi/N)kn}$$

where W_N substitutes for $e^{-j(2\pi/N)}$. Algorithms in which the decomposition is based on splitting the sequence $x(n)$ into smaller sequences are called decimation in time algorithms. The principle of decimation in time is presented below for N equal to an integer power of 2. We can consider, in this case, $X(k)$ to be formed by two

$N/2$-point sequences consisting of the even-numbered points in $x(n)$ and odd-numbered points in $x(n)$, respectively. Thus we obtain

$$X(k) = \sum_{n = 2p + 1} x(n) \, W_N^{nk} + \sum_{n = 2p} x(n) \, W_N^{nk} \tag{11.7}$$

which can also be written as

$$X(k) = \sum_{p = 0}^{N/2 - 1} x(2p) \, W_N^{2pk} + \sum_{p = 0}^{N/2 - 1} x\,(2p + 1) \, W_N^{(2p + 1)k} \tag{11.8}$$

But $W_N^2 = W_{N/2}$ and consequently Eq. (11.8) can be written as

$$X(k) = \sum_{p = 0}^{N/2 - 1} x(2p) W_{N/2}^{pk} + W_N^k \sum_{p = 0}^{N/2 - 1} x(2p + 1) W_{N/2}^{pk} = X_e(k) + W_N^k \, X_o(k) \tag{11.9}$$

Each of the sums in Eq. (11.9) is an $N/2$-point DFT of the even- and odd-numbered points of the original sequence, respectively.

After the two DFTs are computed, they are combined to give the DFT for the original N-point sequence. We can proceed further by decomposing each of the two $N/2$-point DFTs into two $N/4$-point DFTs and each of the four $N/4$-point DFTs into two $N/8$-point DFTs, and so forth. Finally we reduce the computation of the N-point DFT to the computation of the 2-point DFTs and the necessary additions and multiplications.

Figure 11.3 shows the computations involved in computing $X(k)$ for an 8-point original sequence. Oppenheim and Schafer (1975) show that the total number of complex additions and multiplications involved is $N\log_2 N$. The original N-point DFT requires N^2 complex multiplications and additions; thus it turns out that the FFT algorithm saves us considerable computing time.

Figure 11.4 shows the computation time for the FFT and the original DFT versus N. The FFT requires at least an order of magnitude fewer computations than the DFT. As an example, some modern microcomputers equipped with a math coprocessor are able to perform an FFT for a 1024-point sequence in much less than 1 s. In the case when N is not an integer power of 2, the common procedure is to augment the finite-length sequence with zeros until the total number of points reaches the closest power of 2, or the power for which the FFT algorithm is written. This technique is called zero padding. In order to make the error as low as possible, sometimes the signal is multiplied with a finite-length window function. Windowing is also applied when N is an integer power of 2 but the FFT-analyzed signal does not contain an integer number of periods within the N points. In such cases, the error introduced by the unfinished period of the signal may be reduced by a proper choice of the window type.

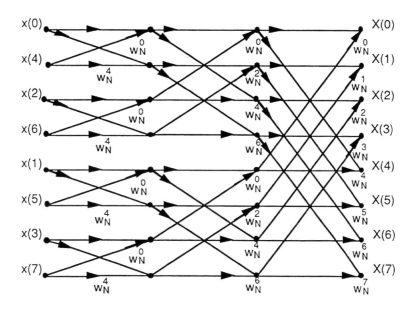

Figure 11.3 The flow graph of the decimation-in-time of an 8-point DFT.

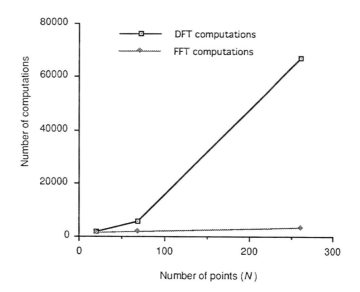

Figure 11.4 Computational savings with the FFT.

11.1.4 C-language FFT function

The computational flow graph presented in Figure 11.3 describes an algorithm for the computation of the FFT of finite-length signal applicable when the number of points is an integer power of 2. When implementing the computations depicted in this figure, we can imagine the use of two arrays of storage registers, one for the array being computed and one for the data being used in computation. For example, in computing the first array, one set of storage registers would contain the input data and the second set of storage registers would contain the computed results for the first stage.

In order to perform the computation based on the "butterfly" graph, the input data must be stored in a nonsequential order. In fact, the order in which the input data must be stored is the bit-reversed order.

To show what is meant by this, we write the index of the output data and the index of the corresponding input data using three binary digits.

$$X(000) — x(000)$$
$$X(001) — x(100)$$
$$X(010) — x(010)$$
$$X(011) — x(110)$$
$$X(100) — x(001)$$
$$X(101) — x(101)$$
$$X(110) — x(011)$$
$$X(111) — x(111)$$

If $(n_2\ n_1\ n_0)$ is the binary representation of the index of the sequence $x(n)$, which is the input data, then $\{x(n)\}$ must be rearranged such that the new position of $x(n_2\ n_1\ n_0)$ is $x(n_0\ n_1\ n_2)$.

Figure 11.5 shows a C-language fragment for the FFT computation, which reorders the input array, $x[nn]$. The FFT function in UW DigiScope allows the user to zero-pad the signal or to window it with the different windows.

11.2 CORRELATION

We now investigate the concept of correlation between groups of data or between signals. Correlation between groups of data implies that they move or change with respect to each other in a structured way. To study the correlation between signals, we will consider signals that have been digitized and that therefore form groups of data.

```
#define RORD(a,b) tempr=(a);(a)=(b);(b)=tempr
...
...
float tempr,x[512];
int i,j,m,n,nn;
...
...
nn=512;
n=nn<<1;
j=1;
for (i=1;i<n;i++){
    if (j>i){
        RORD(x[j],x[i]);/*this is the bit-reversal section of*/
        RORD(x[j+1],x[i+1]);/*a FFT computation routine*/
    }
    m=n >>1;
    while (m>=2 && j>m){
        j-=m;
        m >>1;
    }
    j+=m;
}
```

Figure 11.5 The C-language program for bit-reversal computations.

11.2.1 Correlation in the time domain

For N pairs of data $\{x(n),y(n)\}$, the correlation coefficient is defined as

$$rxy = \frac{\sum\limits_{n=1}^{N} \{x(n) - \bar{x}\} \{y(n) - \bar{y}\}}{\sqrt{\sum\limits_{n=1}^{N} \{x(n) - \bar{x}\}^2 \sum\limits_{n=1}^{N} \{y(n) - \bar{y}\}^2}} \qquad (11.10)$$

If finite-length signals are to be analyzed, then we must define the crosscorrelation function of the two signals.

$$r_{xy}(k) = \frac{\sum\limits_{n=1}^{N} \{x(n) - \bar{x}\} \{y(n + k) - \bar{y}\}}{\sqrt{\sum\limits_{n=1}^{N} \{x(n) - \bar{x}\}^2 \sum\limits_{n=1}^{N} \{y(n) - \bar{y}\}^2}} \qquad (11.11)$$

In the case when the two input signals are the same, the crosscorrelation function becomes the autocorrelation function of that signal. Thus, the autocorrelation function is defined as

$$r_{xx}(k) = \frac{\sum\limits_{n=1}^{N} \{x(n) - \bar{x}\} \{x(n+k) - \bar{x}\}}{\sum\limits_{n=1}^{N} \{x(n) - \bar{x}\}^2} \quad (11.12)$$

Figure 11.6 presents the crosscorrelation of respiratory signals recorded simultaneously from a human subject using impedance pneumography. In Figure 11.6(a), the signals were acquired at different points along the midaxillary line of the subject. The subject was breathing regularly at the beginning of the recording, moving without breathing in the middle of the recording, and breathing regularly again at the end. In Figure 11.6(b), each combination of two recorded channels were crosscorrelated in order to try to differentiate between movement and regular breathing.

11.2.2 Correlation in the frequency domain

The original definition for the crosscorrelation was for continuous signals. Thus if $h(t)$ and $g(t)$ are two continuous signals, then their crosscorrelation function is defined as

$$c_{gh}(t) = \int_{-\infty}^{\infty} g(\tau)\, h(t+\tau)d\tau \quad (11.13)$$

The Fourier transform of the crosscorrelation function satisfies

$$Corr\,(\omega) = G(\omega)^*\, H(\omega) \quad (11.14)$$

where $G(\omega)^*$ is the complex conjugate of $G(\omega)$.

Thus if we consider h and g to be digitized, we may approximate the crosscorrelation function as

$$c_{gh}(m) = \frac{1}{N} \sum_{n=0}^{N-1} g(n)\, h(m+n) \quad (11.15)$$

This equation is also known as the biased estimator of the crosscorrelation function. Between the DFTs of the two input discrete signals and the DFT of the biased estimator, we have the relationship

$$Corr\,(k) = G(k)^*\, H(k) \quad (11.16)$$

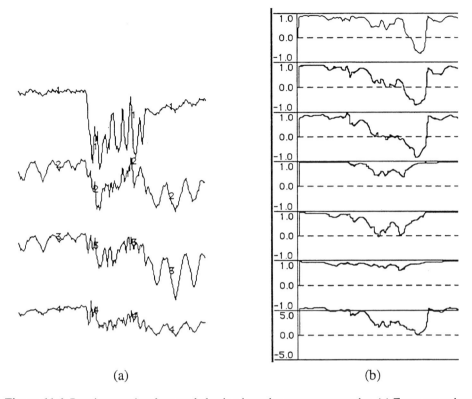

(a)   (b)

Figure 11.6 Respiratory signals recorded using impedance pneumography. (a) From top to bottom, signals 1, 2, 3, and 4 were simultaneously recorded along the midaxillary line of a human subject using a sampling rate of 5 sps. (b) From top to bottom, the traces represent the crosscorrelation of the channels 1–2, 1–3, 1–4, 2–3, 2–4, 3–4, and the averaged correlation coefficient. The results show, in part, that channels 2 and 3 are highly correlated during normal breathing without motion, but not well correlated during motion without breathing.

Thus, the crosscorrelation of the two discrete signals can also be computed as the inverse DFT of the product of their DFTs. This can be implemented using the FFT and inverse FFT (IFFT) algorithms for increasing the computational speed in the way given by

$$c_{gh}(n) = \text{IFFT} \ (\ \text{FFT}^*(g) \ * \ \text{FFT}(h)) \tag{11.17}$$

11.2.3 Correlation function

The C-language program in Figure 11.7 computes the crosscorrelation function of two 512-point input sequences $x[512]$ and $y[512]$ and stores the output data into $rxy[512]$. The idea of this program was used to implement the C-language function to compute the crosscorrelation between an ECG signal and a template. In such a case, the array y[ ] must have the same dimension as the template.

```
/* The crosscorrelation function of x[ ] and y[ ] is */
/* output into rxy[ ] */

void corr(float *x,float *y)
{
  int i,m,n;
  float s,s1,s2,xm,ym,t;
  float rxy[512];
  n=512;
  s=s1=s2=xm=ym=0.0;
  for (i=0;i<n;i++){
      xm=xm+x[i];
      ym=ym+y[i];   /* the arithmetic mean of x[ ] and y[ ] */
  }                 /* computed */
  xm=xm/(float)n;
  ym=ym/(float)n;
    for   ( i=0 ; i<n ; i++){
              s1=s1+pow((x[i]-xm),2.0);
              s2=s2+pow((y[i]-ym),2.0);
    }
    s=sqrt(s1*s2);
    for ( m=0 ; m<n ; m++) {
              t=0.0;
              for ( i=0 ;i<n ;i++) {
                      t=t+(x[i]-xm)*(y[(i+m)%n]-ym);
              }
              rxy[m]=t/s;
    }
}
```

Figure 11.7 C-language function for computing crosscorrelation.

11.3 CONVOLUTION

11.3.1 Convolution in the time domain

It is well known that the passage of a signal through a linear system can be described in the frequency domain by the frequency response of the system. In the time domain, the response of the system to a specific input signal can be described using convolution. Thus, the convolution is the time-domain operation, which is the equivalent of the process of multiplication of the frequency spectra, of the input signal, and of the pulse response of the analyzed system, in order to find the frequency-domain description of the output signal.

If a continuous signal $x(t)$ is applied at the input of an analogous system which has the impulse response $h(t)$, then the equation for the output signal $y(t)$ is known as the convolution equation.

$$y(t) = \int_{-\infty}^{\infty} x(t - \tau)\, h(\tau)\, d\tau \qquad (11.18)$$

For discrete signals, the convolution equation becomes

$$y(m) = \sum_{n=0}^{N-1} x(m - n)\, h(n) \qquad (11.19)$$

where $x(n)$ is the input signal, $h(n)$ is the sampled impulse response of the system, and $y(n)$ is the output signal. Equation (11.19) can be used for implementing finite impulse response digital filters. In this case, $h(n)$ would be a finite-length sequence which represents, in fact, the coefficients of the FIR filter.

Figure 11.8 shows the input and output signals for a low-pass FIR filter with nine coefficients used for filtering the cardiogenic artifact from respiratory signals recorded using impedance pneumography. The corner frequency of this filter is 0.7 Hz, and the attenuation in the stopband is about 20 dB.

11.3.2 Convolution in the frequency domain

Time-domain convolution is often expressed in a shorthand notation using a '$*$' operator, thus

$$y(t) = x(t) * h(t) = h(t) * x(t)$$

or for discrete signals

$$y(n) = x(n) * h(n) = h(n)*x(n)$$

The following relates the DFTs:

$$Y(k) = X(k)\, H(k)$$

If the FFT is used for computing time-domain convolution, this method is called "fast convolution." We obtain

$$x(n) *h(n) = \text{IFFT}(X(k)\, H(k)) \qquad (11.20)$$

where $X(k)$ and $H(k)$ are computed using the FFT algorithm.

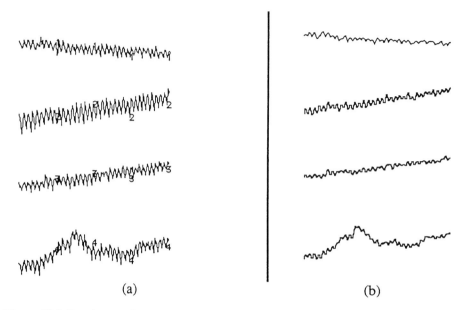

Figure 11.8 Respiratory signals recorded using impedance pneumography. (a) The signals from four different recording channels sampled at a rate of 5 sps. (b) The corresponding signals after being filtered with a 9-coefficient low-pass filter. The cardiogenic artifact, represented by the additive noise seen in (a), is much attenuated by the filtering process.

Convolution in the frequency domain has a similar definition to that for convolution in the time domain. For continuous spectra it is expressed by the integral

$$Y(\omega) = \frac{1}{2\pi} \int_{-\infty}^{\infty} X(\omega - \Omega) \, S(\Omega) \, d\Omega \qquad (11.21)$$

The time-domain equivalent for $Y(\omega)$ is

$$y(t) = x(t) \, s(t)$$

Thus, the multiplication of two signals in the time domain is equivalent to the convolution of their Fourier transforms in the frequency domain.

The same principles apply to discrete signals. We know that the discrete signal spectrum has a basic structure defined in the interval $-f_S/2 \leq f \leq +f_S/2$, where f_S is the sampling frequency. Outside this interval, the spectrum of the sampled signal repeats identically, in positive and negative frequencies. If two discrete signals are multiplied together in the time domain the resulting frequency spectrum would also repeat identically at intervals of sampling frequency. The repeating function would of course be the convolution of the Fourier transforms of the two sampled signals. It is important to note that its form would not be identical with the form of the spectrum of the convolution of two continuous time signals that had the shapes of the envelopes of the sampled signals under consideration.

The concept of convolution in the frequency domain is fundamental to the signal windowing approach. As an example, if $h(n)$ is the impulse response of an ideal low-pass filter with frequency characteristics $H(\omega)$, $h(n)$ will be an infinite length sequence. For example, in order to implement an FIR filter which approximates $H(\omega)$, we must window $h(n)$. Thus, we obtain $h'(n)$ given by

$$h'(n) = w(n)\, h(n)$$ (11.22)

where $w(n)$ is the finite-length windowing sequence. We can obtain the Fourier transform of the implemented FIR filter, $H'(\omega)$, using convolution in the frequency domain. Thus, we get

$$H'(\omega) = H(\omega) * W(\omega)$$ (11.23)

where $W(\omega)$ is the Fourier transform of the windowing sequence. How well $h(n)$ is approximated by $h'(n)$ depends on the windowing sequence properties.

One way to analyze the performance of the window is to study its Fourier transform. In this approach, one may be interested in analyzing the attenuation in the stopband and the transition width. Figure 11.9 presents the most important parameters, which are mostly used in low-pass filter design, for several types of windows. The designer should make a trade-off between the transition width, the number of coefficients, and the minimum attenuation in the stopband. As shown in Figure 11.9, for middle values of the transition width, the best results are obtained using a Hamming window. If the designer is not interested in the transition width performance, then the best results are obtained using a Blackman window. For a detailed approach of the windowing theory, see Oppenheim and Schafer (1975).

Window	Transition width of the main lobe	Minimum stopband attenuation
Rectangular	2fs/N	−21 dB
Hanning	4fs/N	−44 dB
Hamming	4fs/N	−53 dB
Blackman	6fs/N	−74 dB

Figure 11.9 The performance of different window functions. N represents the number of coefficients used in the function which describes the window, and fs is the sampling frequency.

11.3.3 Convolution function

Figure 11.10 gives a C-language function that computes convolution in the time domain between an input 512-point data sequence $x[512]$ and $h[ncoef]$, which might be the coefficient array of an FIR filter, and stores the output data into $cxy[512]$.

```
/* The convolution between x[ ] and h[ ] is saved into cxy[ ] */

conv(float *x,float *h,int ncoef)
{
  int i,m,n;
  float cxy[512];
  n=512;
    for ( m=0 ; m<n ; m++) {
            cxy[m]=0.0;
            for ( i=0 ;i<ncoef && i<=m ;i++) {
                    cxy[m]+= h[i]*x[m-i];
            }
    }
}
```

Figure 11.10 C-language function for computing convolution.

11.4 POWER SPECTRUM ESTIMATION

11.4.1 Parseval's theorem

Parseval's theorem expresses the conservation of energy principle between the time and frequency domains. For a periodic signal $f(t)$ with the period T, Parseval's theorem tells us how to compute the average power contained in this signal knowing the Fourier series coefficients a_k and b_k, $k = 0, ..., \infty$

$$\frac{1}{T} \int_{-T/2}^{T/2} f^2(t)\, dt = a_0^2 + \sum_{k=1}^{\infty} \left(\frac{a_k^2}{2} + \frac{b_k^2}{2} \right) \qquad (11.24)$$

For a continuous aperiodic signal, we have a similar relationship between $f(t)$ and its Fourier pair

$$\int_{-\infty}^{\infty} |f(t)|^2\, dt = \frac{1}{2\pi} \int_{-\infty}^{\infty} |F(\omega)|^2\, d\omega \qquad (11.25)$$

Similarly, for the Fourier transform of real discrete signals

$$\sum_{-\infty}^{\infty} f^2(n) = \frac{1}{2\pi} \int_{2\pi} |F(\Omega)|^2 \, d\Omega \qquad (11.26)$$

In the case of the DFT we assume that the time-domain signal repeats identically with a period of N points, thus the DFT will repeat at intervals of sampling frequency. Parseval's theorem is expressed under these conditions as

$$\sum_{n=0}^{N-1} x^2(n) = \frac{1}{N} \sum_{k=0}^{N-1} |X(k)|^2 \qquad (11.27)$$

To estimate the average power of the signal, we compute the mean squared amplitude and make the approximation

$$\frac{1}{T} \int_{-T/2}^{T/2} f^2(t) \, dt \approx \frac{1}{N} \sum_{n=0}^{N-1} f^2(n) \qquad (11.28)$$

The method for power spectrum estimation (PSE) used in this section is based on the periodogram concept. Thus, if we sample a function $c(t)$ and use the FFT to compute its DFT, we get

$$C_k = \sum_{n=0}^{N-1} c(n) \, e^{\frac{-jkn \, 2\pi}{N}}$$

Then the periodogram estimate for the power spectrum is defined for $N/2 + 1$ frequencies as

$$P(0) = \frac{1}{N^2} |C_0|^2$$

$$P(k) = \frac{1}{N^2} (|C_k|^2 + |C_{N-k}|^2) \quad \text{for } k = 1, \ldots, N/2 - 1 \qquad (11.29)$$

$$P(N/2) = \frac{1}{N^2} |C_{N/2}|^2$$

By Parseval's theorem, we see that the mean squared amplitude is equal to the sum of the $N/2 + 1$ values of P. We must ask ourselves how accurate this estimator is and how it can be improved. The following sections provide two methods for improving the performance of the estimator.

11.4.2 Welch's method of averaging modified periodograms

The periodogram is not a consistent spectral estimator. The variance of the estimate does not tend to go to zero as the record length approaches infinity. One method for improving the estimator proposed by Welch is based on breaking up the N-point data record $x(n)$ into M-point segments $x_k(n)$ that overlap with each other by L samples. If $L = M$ then $N = (K + 1) M$ where K is the total number of segments. Subsequently a window function is applied to each segment. Then a periodogram is computed for each windowed segment. Finally, these periodograms are averaged, and the result is scaled to obtain the Welch estimate.

11.4.3 Blackman-Tukey spectral estimate

The Blackman-Tukey estimation method can be implemented in three steps. In the first step, the middle $2M + 1$ samples of the autocorrelation sequence, $\emptyset_{xx}(m)$, where $-M \leq m \leq M$, are estimated from the available N-point data record. The second step is to apply a window to the estimated autocorrelation lags. Finally, the FFT is computed for the windowed autocorrelation estimate to yield the Blackman-Tukey estimate. The parameter M and the window type must be selected in accordance with the specific application.

11.4.4 Compressed spectral array and gray-scale plots

In the compressed spectral array (CSA) method, the resulting spectra are plotted in time sequence (each power spectrum is plotted slightly above the previous spectrum) in order to produce a three-dimensional effect, so that the resultant plots can be easily interpreted. To show this effect on a two-dimensional graphics printout, each subsequent power spectrum, representing a successive time period, is plotted with its origin shifted in both the x and y directions. The more the origins are shifted in the y direction relative to the x direction, the sharper the viewing angle, which allows for better separation of the individual spectra but makes for a greater difficulty in following a frequency component through several time periods. Figure 11.11(a) shows the CSA as a result of spectral analysis of the EGG (electrogastrogram) of a diabetic patient whose time-domain record shows a tachyarrhythmia (Pfister et al., 1988).

Figure 11.11(b) shows the corresponding gray-scale plot. This is a two-dimensional plot with the x axis representing frequency, the y axis representing time, and the intensity of the points representing the spectral power. The darker a point, the greater the spectral power at that point. Each data point in the gray-scale plot is represented by a 5 X 5 matrix of pixels. Each matrix data point can have one of 26 intensity levels, from all pixels off to all pixels on. All other values are scaled proportionally to the maximal level and rounded to an integer value that represents the intensity level. The gray-scale plot does not provide as great a degree of resolution of amplitude as does the CSA method, but it does facilitate observation of frequency shifts.

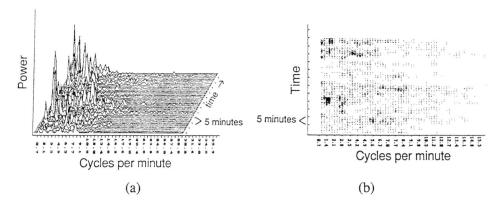

Figure 11.11 Electrogastrogram (EGG) of a diabetic patient. (a) Compressed spectral array (CSA). (b) Gray-scale plot.

11.4.5 Power spectrum function

Figure 11.12 is a C-language function that computes the power spectrum for an N-point data sequence ($N = 512$). The input data are taken from an input buffer as integers, converted to floating-point format, and then used to compute the power spectrum. The output data are scaled and saved in an output file as integers after conversion from the floating-point format. The program presented is based on Welch's idea of periodogram averaging. Thus, this function divides the input sequence of data into two 256-point segments, windows each segment, and performs the power spectrum estimation for each segment. Finally, the results obtained for each segment are averaged and scaled according to the window effects.

11.5 LAB: FREQUENCY-DOMAIN ANALYSIS OF THE ECG

This lab provides experience in studying the frequency characteristics of signals. Another resource for practice with these techniques is Alkin (1991).

11.5.1 Power spectral analysis of periodic signals

Use **(G)enwave** to create sine wave and square wave signals at several different fundamental frequencies using several different sampling rates. Use the **(P)wr Spect** command to compute and observe the frequency spectrum of each signal and comment on how sampling rate affects the results. How could you use a power spectrum routine to obtain the actual Fourier coefficients?

11.5.2 Power spectral analysis of an ECG

1. Use the (P)wr Spect command to find the frequency corresponding to the main peak in the frequency spectrum of an ECG. Is the result what you expected?

2. Select ad(V) Ops from the main menu, and choose a QRS complex using (T)emplate. Use the (P)wr Spect command to find the spectrum of the zero-padded QRS complex. Is the frequency corresponding to the main peak in the frequency spectrum of the QRS complex what you expected?

3. After selecting a QRS complex template, use the (W)indow command to window the selected template, and then run the (P)wr Spect command. Document the effects of the various windows.

4. Select a P-QRS-T segment as a template. Using the (P)wr Spect command, find the power spectrum estimate for an ECG. Find the frequency corresponding to the main peak in the frequency spectrum of an ECG. How does it differ from the results of part 1?

```
#define WINDOW(j,a,b)(1.0-fabs((((j)-1)-(a))*(b)))/*parzen */
#define SQR(a)  (sqrarg=(a),sqrarg*sqrarg)/*modulus of a,squared*/
...
  for(j=1;j<=mm;j++) {
          w=WINDOW(j,facm,facp);
              w1[2*j-2] *=w; /* the real part of then data segment
                          is windowed */
              w1[2*j-1]  =0.0; /*the imaginary part is
                                  set to zero */
  }
...
fft(w1-1,mm,1);   /*the fft of the windowed signal is performed*/
/* the power spectrum estimate for the windowed data segment
       is computed */
p[0] +=(SQR(w1[0]+SQR(w1[1]));
for (j=1;j<m;j++) {
          j2=2*j;
          p[j] +=(SQR(w1[j2])+SQR(w1[j2+1])+SQR(w1[-j2+4*m])+
                  SQR(w1[-j2+4*m+1])));
  }
p[m] +=(SQR(w1[mm]+sqr(w1[mm+1])));
...
for (j=0;j<=m;j++)  p[j] /= scale;         /*the PSE is scaled */
                              /* considering the window effects */
```

Figure 11.12 C-language fragment for computing a power spectrum estimation.

11.5.3 Crosscorrelation of the ECG

Select a QRS complex using the (T)emplate command. Use the (C)orrelation command to correlate this QRS with the ECG. Explain the relationship between the output of the crosscorrelation function and the size of the selected template. What is the time delay between the peak of the crosscorrelation function and the selected QRS? Read a different ECG from a disk file, and crosscorrelate the template with the new ECG. Explain all your observations.

11.6 REFERENCES

Alkin, O. 1991. *PC-DSP*, Englewood Cliffs, NJ: Prentice Hall.

Oppenheim, A. V., and Schafer, R. W. 1975. *Digital Signal Processing*. Englewood Cliffs, NJ: Prentice Hall.

Pfister, C. J., Hamilton, J. W., Bass, P., Webster, J. G., and Tompkins, W. J. 1988. Use of spectral analysis in detection of frequency differences in electrogastrograms of normal and diabetic subjects. *IEEE Trans. Biomed. Eng.* 935–41.

11.7 STUDY QUESTIONS

11.1 Derive Eqs. (11.24) and (11.27) considering $x(n)$ to be a digital periodic signal containing N samples in a period.

11.2 If $x(n) = 1.0 + \cos(2\pi n/4.0)$, find the DFT of this signal for $n = 0, \ldots, 7$. Write a C-language program that creates a data file with the values of this signal. Read this data file with the UW DigiScope program, and pad it with the corresponding number of zeros. Take the power spectrum of the zero-padded signal using the (P)wr Spect command without windowing it, and compare the result with your hand analysis. Explain the differences.

11.3 Create a file that has the values of $x(n) = \sin(2\pi n/512.0)$, $n = 0, \ldots, 511$. Take the power spectrum of this signal without windowing it. Replace the last 10 samples of this signal with zeros. Take the power spectrum of the padded signal with and without a window. Explain the differences.

11.4 Select one period of the signal created in question 11.2 as a template. Crosscorrelate this template with the created signal. Explain the shape of the crosscorrelation function. Compute the amplitude of the peak and compare it with the result of your experiment.

11.5 A 100-Hz-bandwidth ECG signal is sampled at a rate of 500 samples/s. (a) Draw the approximate frequency spectrum of the new digital signal obtained after sampling, and label important points on the axes. (b) On the same graph, draw the approximate spectrum that would be averaged from a set of normal QRS complexes.

12

ECG QRS Detection

Valtino X. Afonso

Over the past few years, there has been an increased trend toward processing of the electrocardiogram (ECG) using microcomputers. A survey of literature in this research area indicates that systems based on microcomputers can perform needed medical services in an extremely efficient manner. In fact, many systems have already been designed and implemented to perform signal processing tasks such as 12-lead off-line ECG analysis, Holter tape analysis, and real-time patient monitoring. All these applications require an accurate detection of the QRS complex of the ECG. For example, arrhythmia monitors for ambulatory patients analyze the ECG in real time (Pan and Tompkins, 1985), and when an arrhythmia occurs, the monitor stores a time segment of the abnormal ECG. This kind of monitor requires an accurate QRS recognition capability. Thus, QRS detection is an important part of many ECG signal processing systems.

This chapter discusses a few of the many techniques that have been developed to detect the QRS complex of the ECG. It begins with a discussion of the power spectrum of the ECG and goes on to review a variety of QRS detection algorithms.

12.1 POWER SPECTRUM OF THE ECG

The power spectrum of the ECG signal can provide useful information about the QRS complex. This section reiterates the notion of the power spectrum presented earlier, but also gives an interpretation of the power spectrum of the QRS complex. The power spectrum (based on the FFT) of a set of 512 sample points that contain approximately two heartbeats results in a series of coefficients with a maximal value near a frequency corresponding to the heart rate.

The heart rate can be determined by multiplying together the normalized frequency and the sampling frequency. We can also get useful information about the frequency spectrum of the QRS complex. In order to obtain this information, the QRS complex of the ECG signal must be selected as a template and zero-padded

236

prior to the power spectrum analysis. The peak of the frequency spectrum obtained corresponds to the peak energy of the QRS complex.

The ECG waveform contains, in addition to the QRS complex, P and T waves, 60-Hz noise from powerline interference, EMG from muscles, motion artifact from the electrode and skin interface, and possibly other interference from electro-surgery equipment in the operating room. Many clinical instruments such as a car-diotachometer and an arrhythmia monitor require accurate real-time QRS detec-tion. It is necessary to extract the signal of interest, the QRS complex, from the other noise sources such as the P and T waves. Figure 12.1 summarizes the relative power spectra of the ECG, QRS complexes, P and T waves, motion artifact, and muscle noise based on our previous research (Thakor et al., 1983).

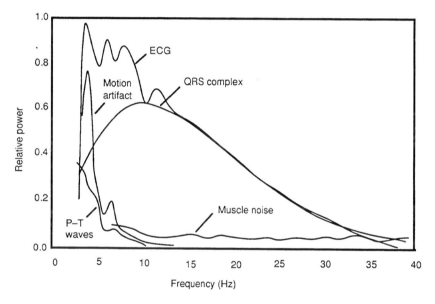

Figure 12.1 Relative power spectra of QRS complex, P and T waves, muscle noise and motion artifacts based on an average of 150 beats.

12.2 BANDPASS FILTERING TECHNIQUES

From the power spectral analysis of the various signal components in the ECG sig-nal, a filter can be designed which effectively selects the QRS complex from the ECG. Another study that we performed examined the spectral plots of the ECG and the QRS complex from 3875 beats (Thakor et al., 1984). Figure 12.2 shows a plot of the signal-to-noise ratio (SNR) as a function of frequency. The study of the power spectra of the ECG signal, QRS complex, and other noises also revealed that a maximum SNR value is obtained for a bandpass filter with a center frequency of 17 Hz and a Q of 3. Section 12.3 and a laboratory experiment examine the effects of different values of the Q of such a filter.

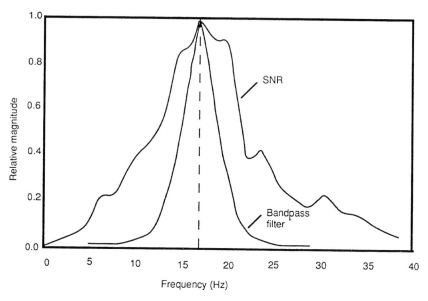

Figure 12.2 Plots of the signal-to-noise ratio (SNR) of the QRS complex referenced to all other signal noise based on 3875 heart beats. The optimal bandpass filter for a cardiotachometer maximizes the SNR.

12.2.1 Two-pole recursive filter

A simple two-pole recursive filter can be implemented in the C language to bandpass the ECG signal. The difference equation for the filter is

$$y(nT) = 1.875y(nT - T) - 0.9219y(nT - 2T) + x\,(nT) - x(nT - 2T) \qquad (12.1)$$

This filter design assumes that the ECG signal is sampled at 500 samples/s. The values of 1.875 and 0.9219 are approximations of the actual design values of 1.87635 and 0.9216 respectively. Since the coefficients are represented as powers of two, the multiplication operations can be implemented relatively fast using the shift operators in the C language. Figure 12.3 displays the code fragment that implements Eq. (12.1).

```
twoPoleRecursive(int data)
{
        static int xnt, xm1, xm2, ynt, ym1, ym2 = 0;
        xnt = data;

        ynt = (ym1 + ym1 >> 1 + ym1 >> 2 + ym1 >> 3) +
              (ym2 >> 1 + ym2 >> 2 + ym2 >> 3 +
               ym2 >> 5 + ym2 >> 6) + xnt - xm2;

        xm2 = xm1;
        xm1 = xnt;
        xm2 = ym1;
        ym2 = ym1;
        ym1 = ynt;
        return(ynt);
}
```

Figure 12.3 C-language code to implement a simple two-pole recursive filter.

Note that in this code, the coefficients 1.87635 and 0.9216 are approximated by

$$1.875 = 1 + \frac{1}{2} + \frac{1}{4} + \frac{1}{8}$$

and

$$0.9219 = \frac{1}{2} + \frac{1}{4} + \frac{1}{8} + \frac{1}{32} + \frac{1}{64}$$

12.2.2 Integer filter

An approximate integer filter can be realized using the general form of the transfer function given in Chapter 7. QRS detectors for cardiotachometer applications frequently bandpass the ECG signal using a center frequency of 17 Hz. The denominator of the general form of the transfer function allows for poles at 60°, 90°, and 120°, and these correspond to center frequencies of a bandpass filter of $T/6$, $T/4$, and $T/3$ Hz, respectively. The desired center frequency can thus be obtained by choosing an appropriate sampling frequency.

Ahlstrom and Tompkins (1985) describe a useful filter for QRS detection that is based on the following transfer function:

$$H(z) = \frac{(1 - z^{-12})^2}{(1 - z^{-1} + z^{-2})^2} \tag{12.2}$$

This filter has 24 zeros at 12 different frequencies on the unit circle with poles at ±60°. The ECG signal is sampled at 200 sps, and then the turning point algorithm is used to reduce the sampling rate to 100 sps. The center frequency is at 16.67 Hz and the nominal bandwidth is ±8.3 Hz. The duration of the ringing is approximately 240 ms (the next section explains the effects of different filter Qs). The difference equation to implement this transfer function is

$$y(nT) = 2y(nT - T) - 3y(nT - 2T) + 2y(nT - 3T) - y(nT - 4T)$$
$$+ x(nT) - 2x(nT - 12T) + x(nT - 24T) \qquad (12.3)$$

12.2.3 Filter responses for different values of Q

The value of Q of the bandpass filter centered at $f_c = 17$ Hz determines how well the signal of interest is passed without being attenuated. It is also necessary to increase the SNR of the signal of interest; that is, the QRS complex. The Q of the filter is calculated as

$$Q = \frac{f_c}{BW} \qquad (12.4)$$

A value of Q that is too high will result in a very oscillatory response (Thakor et al., 1984). The ripples must die down within 200 ms. This is necessary so that the ripples from one QRS complex do not interfere with the ripples from the next one. With a center frequency of 17 Hz, the maximal permissible Q was found to be 5. Figure 12.4 shows the effect of different values of Q. For a bandpass filter with $f_c = 17$ Hz, a Q value of 5 was found to maximize the SNR (Thakor et al., 1984).

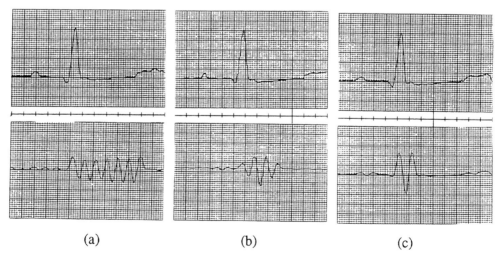

(a)   (b)   (c)

Figure 12.4 Effects of different values of Q. A higher Q results in a oscillatory transient response. (a) $Q = 8$. (b) $Q = 3$. (c) $Q = 1$.

12.3 DIFFERENTIATION TECHNIQUES

Differentiation forms the basis of many QRS detection algorithms. Since it is basically a high-pass filter, the derivative amplifies the higher frequencies characteristic of the QRS complex while attenuating the lower frequencies of the P and T waves.

An algorithm based on first and second derivatives originally developed by Balda et al. (1977) was modified for use in high-speed analysis of recorded ECGs by Ahlstrom and Tompkins (1983). Friesen et al. (1990) subsequently implemented the algorithm as part of a study to compare noise sensitivity among certain types of QRS detection algorithms. Figure 12.5 shows the signal processing steps of this algorithm.

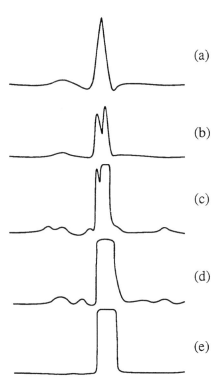

(a)

(b)

(c)

(d)

(e)

Figure 12.5 Various signal stages in the QRS detection algorithm based on differentiation. (a) Original ECG. (b) Smoothed and rectified first derivative. (c) Smoothed and rectified second derivative. (d) Smoothed sum of (b) and (c). (e) Square pulse output for each QRS complex.

The absolute values of the first and second derivative are calculated from the ECG signal

$$y0(nT) = |x(nT) - x(nT - 2T)| \qquad (12.5)$$

$$y_1(nT) = |x(nT) - 2x(nT - 2T) + x(nT - 4T)| \qquad (12.6)$$

These two data buffers, $y_0(nT)$ and $y_1(nT)$, are scaled and then summed

$$y_2(nT) = 1.3y_0(nT) + 1.1y_1(nT) \qquad (12.7)$$

The data buffer $y_2(nT)$ is now scanned until a certain threshold is met or exceeded

$$y_2(iT) \geq 1.0 \qquad (12.8)$$

Once this condition is met for a data point in $y_2(iT)$, the next eight points are compared to the threshold. If six or more of these eight points meet or exceed the threshold, then the segment might be part of the QRS complex. In addition to detecting the QRS complex, this algorithm has the advantage that it produces a pulse which is proportional in width to the complex. However, a disadvantage is that it is particularly sensitive to higher-frequency noise.

12.4 TEMPLATE MATCHING TECHNIQUES

In this section we discuss techniques for classifying patterns in the ECG signal that are quite related to the human recognition process.

12.4.1 Template crosscorrelation

Signals are said to be correlated if the shapes of the waveforms of two signals match one another. The correlation coefficient is a value that determines the degree of match between the shapes of two or more signals. A QRS detection technique designed by Dobbs et al. (1984) uses crosscorrelation.

This technique of correlating one signal with another requires that the two signals be aligned with one another. In this QRS detection technique, the template of the signal that we are trying to match stores a digitized form of the signal shape that we wish to detect. Since the template has to be correlated with the incoming signal, the signal should be aligned with the template. Dobbs et al. describe two ways of implementing this.

The first way of aligning the template and the incoming signal is by using the fiducial points on each signal. These fiducial points have to be assigned to the signal by some external process. If the fiducial points on the template and the signal are aligned, then the correlation can be performed.

Another implementation involves continuous correlation between a segment of the incoming signal and the template. Whenever a new signal data point arrives, the oldest data point in time is discarded from the segment (a first-in-first-out data structure). A correlation is performed between this signal segment and the template segment that has the same number of signal points. This technique does not require processing time to assign fiducial points to the signal. The template can be thought of as a window that moves over the incoming signal one data point at a time. Thus, alignment of the segment of the signal of interest must occur at least once as the window moves through the signal.

The value of the crosscorrelation coefficient always falls between +1 and –1. A value of +1 indicates that the signal and the template match exactly. As mentioned earlier, the value of this coefficient determines how well the *shapes* of the two waveforms under consideration match. The magnitude of the actual signal samples does not matter. This shape matching, or recognizing process of QRS complexes, conforms with our natural approach to recognizing signals.

12.4.2 Template subtraction

Figure 12.6 illustrates a template subtraction technique. This is a relatively simple QRS detection technique as compared to the other ones described in this chapter.

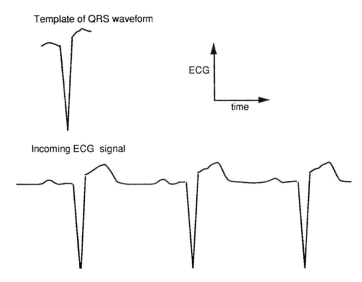

Figure 12.6 In simple template matching, the incoming signal is subtracted, point by point, from the QRS template. If the two waveforms are perfectly aligned, the subtraction results in a zero value.

The algorithm begins by saving a segment of the incoming ECG signal that corresponds to the QRS waveform. This segment or template is then compared with the incoming ECG signal. Each point in the incoming signal is subtracted from the corresponding point in the template. When the template is aligned with a QRS waveform in the signal, the subtraction results in a value very close to zero. This algorithm uses only as many subtraction operations as there are points in the template.

12.4.3 Automata-based template matching

Furno and Tompkins (1982) developed a QRS detector that is based on concepts from automata theory. The algorithm uses some of the basic techniques that are common in many pattern recognition systems. The ECG signal is first reduced into a set of predefined tokens, which represent certain shapes of the ECG waveform.

Figure 12.7 shows the set of tokens that would represent a normal ECG. Then this set of tokens is input to the finite state automaton defined in Figure 12.8. The finite state automaton is essentially a state-transition diagram that can be implemented with IF ... THEN control statements available in most programming languages. The sequence of tokens is fed into the automaton. For example, a sequence of tokens such as *zero, normup, normdown*, and *normup* would result in the automaton signaling a normal classification for the ECG.

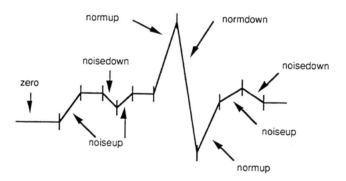

Figure 12.7 Reduction of an ECG signal to tokens.

The sequence of tokens must be derived from the ECG signal data. This is done by forming a sequence of the differences of the input data. Then the algorithm groups together those differences that have the same sign and also exceed a certain predetermined threshold level. The algorithm then sums the differences in each of the groups and associates with each group this sum and the number of differences that are in it.

This QRS detector has an initial learning phase where the program approximately determines the peak magnitude of a normal QRS complex. Then the algorithm detects a normal QRS complex each time there is a deflection in the waveform with a magnitude greater than half of the previously determined peak. The algorithm now teaches the finite state automaton the sequence of tokens that make up a normal QRS complex. The number and sum values (discussed in the preceding paragraph) for a normal QRS complex are now set to a certain range of their respective values in the QRS complex detected.

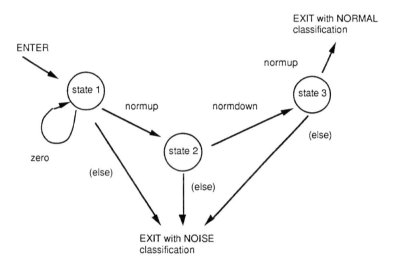

Figure 12.8 State-transition diagram for a simple automaton detecting only normal QRS complexes and noise. The state transition (else) refers to any other token not labeled on a state transition leaving a particular state.

The algorithm can now assign a waveform token to each of the groups formed previously based on the values of the number and the sum in each group of differences. For example, if a particular group of differences has a sum and number value in the ranges (determined in the learning phase) of a QRS upward or downward deflection, then a *normup* or *normdown* token is generated for that group of differences. If the number and sum values do not fall in this range, then a *noiseup* or *noisedown* token is generated. A *zero* token is generated if the sum for a group of differences is zero. Thus, the algorithm reduces the ECG signal data into a sequence of tokens, which can be fed to the finite state automata for QRS detection.

12.5 A QRS DETECTION ALGORITHM

A real-time QRS detection algorithm developed by Pan and Tompkins (1985) was further described by Hamilton and Tompkins (1986). It recognizes QRS complexes based on analyses of the slope, amplitude, and width.

Figure 12.9 shows the various filters involved in the analysis of the ECG signal. In order to attenuate noise, the signal is passed through a bandpass filter composed of cascaded high-pass and low-pass integer filters. Subsequent processes are differentiation, squaring, and time averaging of the signal.

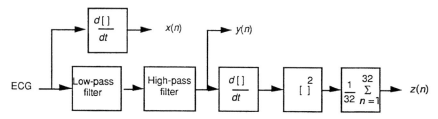

Figure 12.9 Filter stages of the QRS detector. $z(n)$ is the time-averaged signal. $y(n)$ is the bandpassed ECG, and $x(n)$ is the differentiated ECG.

We designed a bandpass filter from a special class of digital filters that require only integer coefficients. This permits the microprocessor to do the signal processing using only integer arithmetic, thereby permitting real-time processing speeds that would be difficult to achieve with floating-point processing. Since it was not possible to directly design the desired bandpass filter with this special approach, the design actually consists of cascaded low-pass and high-pass filter sections. This filter isolates the predominant QRS energy centered at 10 Hz, attenuates the low frequencies characteristic of P and T waves and baseline drift, and also attenuates the higher frequencies associated with electromyographic noise and power line interference.

The next processing step is differentiation, a standard technique for finding the high slopes that normally distinguish the QRS complexes from other ECG waves. To this point in the algorithm, all the processes are accomplished by linear digital filters.

Next is a nonlinear transformation that consists of point-by-point squaring of the signal samples. This transformation serves to make all the data positive prior to subsequent integration, and also accentuates the higher frequencies in the signal obtained from the differentiation process. These higher frequencies are normally characteristic of the QRS complex.

The squared waveform passes through a moving window integrator. This integrator sums the area under the squared waveform over a 150-ms interval, advances one sample interval, and integrates the new 150-ms window. We chose the window's width to be long enough to include the time duration of extended abnormal QRS complexes, but short enough so that it does not overlap both a QRS complex and a T wave.

Adaptive amplitude thresholds applied to the bandpass-filtered waveform and to the moving integration waveform are based on continuously updated estimates of the peak signal level and the peak noise. After preliminary detection by the adaptive thresholds, decision processes make the final determination as to whether or not a detected event was a QRS complex.

A measurement algorithm calculates the QRS duration as each QRS complex is detected. Thus, two waveform features are available for subsequent arrhythmia analysis—RR interval and QRS duration.

Each of the stages in this QRS detection technique are explained in the following sections. Figure 12.10 is a sampled ECG that will serve as an example input signal for the processing steps to follow.

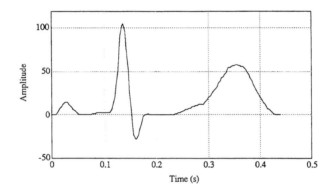

Figure 12.10 Electrocardiogram sampled at 200 samples per second.

12.5.1 Bandpass integer filter

The bandpass filter for the QRS detection algorithm reduces noise in the ECG signal by matching the spectrum of the average QRS complex. Thus, it attenuates noise due to muscle noise, 60-Hz interference, baseline wander, and T-wave interference. The passband that maximizes the QRS energy is approximately in the 5–15 Hz range, as discussed in section 12.1. The filter implemented in this algorithm is a recursive integer filter in which poles are located to cancel the zeros on the unit circle of the *z* plane. A low-pass and a high-pass filter are cascaded to form the bandpass filter.

Low-pass filter

The transfer function of the second-order low-pass filter is

$$H(z) = \frac{(1 - z^{-6})^2}{(1 - z^{-1})^2}$$

(12.9)

The difference equation of this filter is

$$y(nT) = 2y(nT - T) - y(nT - 2T) + x(nT) - 2x(nT - 6T) + x(nT - 12T)$$ (12.10)

The cutoff frequency is about 11 Hz, the delay is five samples (or 25 ms for a sampling rate of 200 sps), and the gain is 36. Figure 12.11 displays the C-language program that implements this low-pass filter. In order to avoid saturation, the output is divided by 32, the closest integer value to the gain of 36 that can be implemented with binary shift arithmetic.

```
int LowPassFilter(int data)
{
    static int y1 = 0, y2 = 0, x[26], n = 12;
    int y0;

    x[n] = x[n + 13] = data;
    y0 = (y1 << 1) - y2 + x[n] - (x[n + 6] << 1) + x[n + 12];
    y2 = y1;
    y1 = y0;
    y0 >>= 5;
    if(--n < 0)
        n = 12;

    return(y0);
}
```

Figure 12.11 C-language program to implement the low-pass filter.

Figure 12.12 shows the performance details of the low-pass filter. This filter has purely linear phase response. Note that there is more than 35-dB attenuation of the frequency corresponding to 0.3 f/f_s. Since the sample rate is 200 sps for these filters, this represents a frequency of 60 Hz. Therefore, power line noise is significantly attenuated by this filter. Also all higher frequencies are attenuated by more than 25 dB.

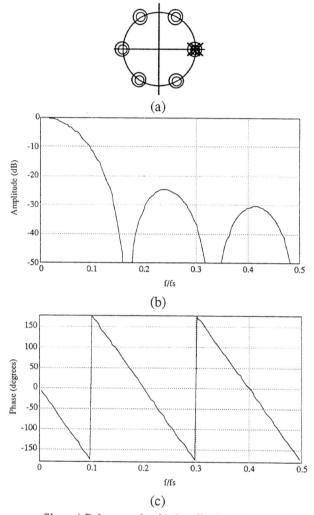

(a)

(b)

(c)

Figure 12.12 Low-pass filter. a) Pole-zero plot. b) Amplitude response. c) Phase response.

Figure 12.13 shows the ECG of Figure 12.10 after processing with the low-pass filter. The most noticeable result is the attenuation of the higher frequency QRS complex. Any 60-Hz noise or muscle noise present would have also been significantly attenuated.

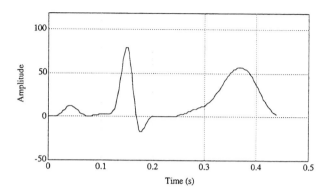

Figure 12.13 Low-pass filtered ECG.

High-pass filter

Figure 12.14 shows how the high-pass filter is implemented by subtracting a first-order low-pass filter from an all-pass filter with delay. The low-pass filter is an integer-coefficient filter with the transfer function

$$H_{lp}(z) = \frac{Y(z)}{X(z)} = \frac{1 - z^{-32}}{1 - z^{-1}} \tag{12.11}$$

and the difference equation

$$y(nT) = y(nT - T) + x(nT) - x(nT - 32T) \tag{12.12}$$

This filter has a dc gain of 32 and a delay of 15.5 samples.

The high-pass filter is obtained by dividing the output of the low-pass filter by its dc gain and then subtracting from the original signal. The transfer function of the high-pass filter is

$$H_{hp}(z) = \frac{P(z)}{X(z)} = z^{-16} - \frac{H_{lp}(z)}{32} \tag{12.13}$$

The difference equation for this filter is

$$p(nT) = x(nT - 16T) - \frac{1}{32} [y(nT - T) + x(nT) - x(nT - 32T)] \tag{12.14}$$

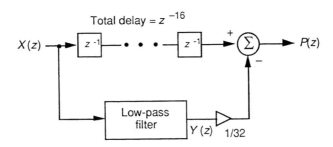

Figure 12.14 The high-pass filter is implemented by subtracting a low-pass filter from an all-pass filter with delay.

The low cutoff frequency of this filter is about 5 Hz, the delay is about $16T$ (or 80 ms), and the gain is 1. Figure 12.15 shows the C-language program that implements this high-pass filter.

```
int HighPassFilter(int data)
{
        static int y1 = 0, x[66], n = 32;
        int y0;

        x[n] = x[n + 33] = data;
        y0 = y1 + x[n] - x[n + 32];
        y1 = y0;
        if(--n < 0)
                n = 32;

        return(x[n + 16] - (y0 >> 5));
}
```

Figure 12.15 C-language program to implement the high-pass filter.

Figure 12.16 shows the performance characteristics of the high-pass filter. Note that this filter also has purely linear phase response.

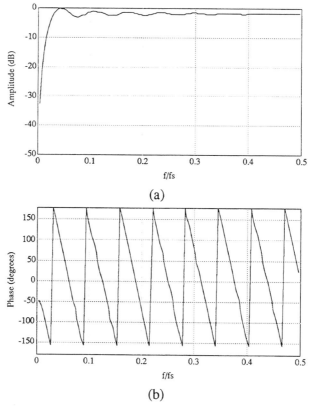

Figure 12.16 High-pass filter. a) Amplitude response. b) Phase response.

Figure 12.17 shows the amplitude response of the bandpass filter which is composed of the cascade of the low-pass and high-pass filters. The center frequency of the passband is at 10 Hz. The amplitude response of this filter is designed to approximate the spectrum of the average QRS complex as illustrated in Figure 12.1. Thus this filter optimally passes the frequencies characteristic of a QRS complex while attenuating lower and higher frequency signals.

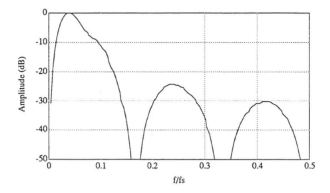

Figure 12.17 Amplitude response of bandpass filter composed of low-pass and high-pass filters.

Figure 12.18 is the resultant signal after the ECG of Figure 12.10 passes through the bandpass filter. Note the attenuation of the T wave due to the high-pass filter.

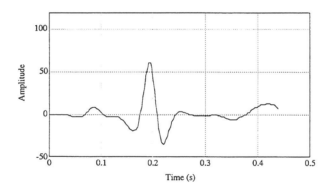

Figure 12.18 Bandpass-filtered ECG.

12.5.2 Derivative

After the signal has been filtered, it is then differentiated to provide information about the slope of the QRS complex. A five-point derivative has the transfer function

$$H(z) = 0.1 \ (2 + z^{-1} - z^{-3} - 2z^{-4}) \tag{12.15}$$

This derivative is implemented with the difference equation

$$y(nT) = \frac{2x(nT) + x(nT - T) - x(nT - 3T) - 2x(nT - 4T)}{8} \qquad (12.16)$$

The fraction 1/8 is an approximation of the actual gain of 0.1. Throughout these filter designs, we approximate parameters with power-of-two values to facilitate real-time operation. These power-of-two calculations are implemented in the C language by shift-left or shift-right operations.

This derivative approximates the ideal derivative in the dc through 30-Hz frequency range. The derivative has a filter delay of $2T$ (or 10 ms). Figure 12.19 shows the C-language program for implementing this derivative.

```
int Derivative(int data)
{
        int y, i;
        static int x_derv[4];

        /*y = 1/8 (2x(nT) + x(nT - T) - x(nT - 3T) - 2x(nT - 4T))*/

        y = (data << 1) + x_derv[3] - x_derv[1] - (x_derv[0] << 1);

        y >>= 3;
        for (i = 0; i < 3; i++)
              x_derv[i] = x_derv[i + 1];
        x_derv[3] = data;

        return(y);
}
```

Figure 12.19 C-language program to implement the derivative.

Figure 12.20 shows the performance characteristics of this derivative implementation. Note that the amplitude response approximates that of a true derivative up to about 20 Hz. This is the important frequency range since all higher frequencies are significantly attenuated by the bandpass filter.

Figure 12.21 is the resultant signal after passing through the cascade of filters including the differentiator. Note that P and T waves are further attenuated while the peak-to-peak signal corresponding to the QRS complex is further enhanced.

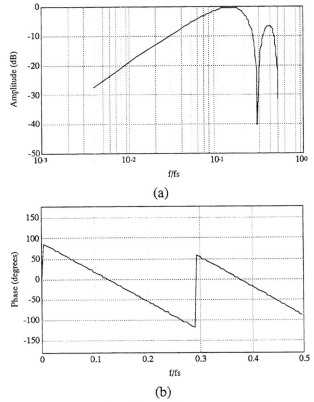

(a)

(b)

Figure 12.20 Derivative. a) Amplitude response. b) Phase response.

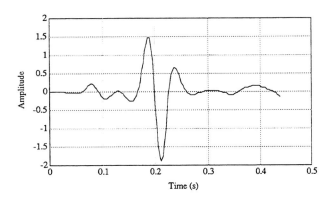

Figure 12.21 ECG after bandpass filtering and differentiation.

12.5.3 Squaring function

The previous processes and the moving-window integration, which is explained in the next section, are linear processing parts of the QRS detector. The squaring function that the signal now passes through is a nonlinear operation. The equation that implements this operation is

$$y(nT) = [x(nT)]^2 \qquad (12.17)$$

This operation makes all data points in the processed signal positive, and it amplifies the output of the derivative process nonlinearly. It emphasizes the higher frequencies in the signal, which are mainly due to the QRS complex. A fact to note in this operation is that the output of this stage should be hardlimited to a certain maximum level corresponding to the number of bits used to represent the data type of the signal. Figure 12.22 shows the results of this processing for our sample ECG.

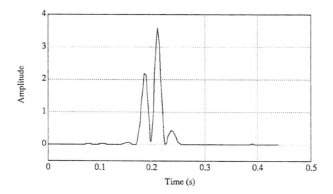

Figure 12.22 ECG signal after squaring function.

12.5.4 Moving window integral

The slope of the R wave alone is not a guaranteed way to detect a QRS event. Many abnormal QRS complexes that have large amplitudes and long durations (not very steep slopes) might not be detected using information about slope of the R wave only. Thus, we need to extract more information from the signal to detect a QRS event.

Moving window integration extracts features in addition to the slope of the R wave. It is implemented with the following difference equation:

$$y(nT) = \frac{1}{N} [x(nT - (N-1)T) + x(nT - (N-2)T) + \ldots + x(nT)] \qquad (12.18)$$

where N is the number of samples in the width of the moving window. The value of this parameter should be chosen carefully.

Figure 12.23 shows the output of the moving window integral for the sample ECG of Figure 12.10. Figure 12.24 illustrates the relationship between the QRS complex and the window width. The width of the window should be approximately the same as the widest possible QRS complex. If the size of the window is too large, the integration waveform will merge the QRS and T complexes together. On the other hand, if the size of the window is too small, a QRS complex could produce several peaks at the output of the stage. The width of the window should be chosen experimentally. For a sample rate of 200 sps, the window chosen for this algorithm was 30 samples wide (which corresponds to 150 ms).

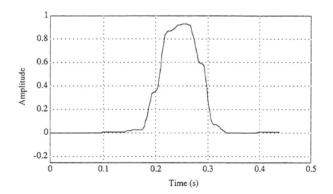

Figure 12.23 Signal after moving window integration.

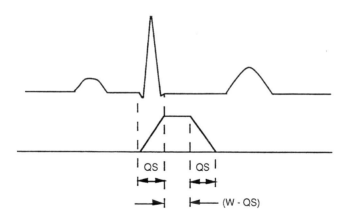

Figure 12.24 The relationship of a QRS complex to the moving integration waveform. (a) ECG signal. (b) Output of moving window integrator. QS: QRS width. W: width of the integrator window.

Figure 12.25 shows the C-language program that implements the moving window integration.

```
int MovingWindowIntegral(int data)
{
        static int x[32], ptr = 0;
        static long sum = 0;
        long ly;
        int y;

        if(++ptr == 32)
                ptr = 0;
        sum -= x[ptr];
        sum += data;
        x[ptr] = data;
        ly = sum >> 5;
        if(ly > 32400)          /*check for register overflow*/
                y = 32400;
        else
                y = (int) ly;

        return(y);
}
```

Figure 12.25 C-language program to implement the moving window integration.

Figure 12.26 shows the appearance of some of the filter outputs of this algorithm. Note the processing delay between the original ECG complexes and corresponding waves in the moving window integral signal.

12.5.5 Thresholding

The set of thresholds that Pan and Tompkins (1985) used for this stage of the QRS detection algorithm were set such that signal peaks (i.e., valid QRS complexes) were detected. Signal peaks are defined as those of the QRS complex, while noise peaks are those of the T waves, muscle noise, etc. After the ECG signal has passed through the bandpass filter stages, its signal-to-noise ratio increases. This permits the use of thresholds that are just above the noise peak levels. Thus, the overall sensitivity of the detector improves.

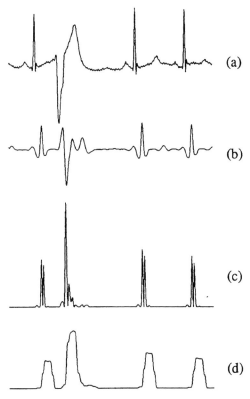

Figure 12.26 QRS detector signals. (a) Unfiltered ECG. (b) Output of bandpass filter. (c) Output after bandpass, differentiation, and squaring processes. (d) Final moving-window integral.

Two sets of thresholds are used, each of which has two threshold levels. The set of thresholds that is applied to the waveform from the moving window integrator is

$$SPKI = 0.125\ PEAKI + 0.875\ SPKI \qquad \text{if } PEAKI \text{ is the signal peak}$$

$$NPKI = 0.125\ PEAKI + 0.875\ NPKI \qquad \text{if } PEAKI \text{ is the noise peak}$$

$$THRESHOLD\ I1 = NPKI + 0.25\ (SPKI - NPKI)$$

$$THRESHOLD\ I2 = 0.5\ THRESHOLD\ I1$$

All the variables in these equations refer to the signal of the integration waveform and are described below:

PEAKI is the overall peak.
SPKI is the running estimate of the signal peak.
NPKI is the running estimate of the noise peak.
THRESHOLD I1 is the first threshold applied.
THRESHOLD I2 is the second threshold applied.

A peak is determined when the signal changes direction within a certain time interval. Thus, *SPKI* is the peak that the algorithm has learned to be that of the QRS complex, while *NPKI* peak is any peak that is not related to the signal of interest. As can be seen from the equations, new values of thresholds are calculated from previous ones, and thus the algorithm adapts to changes in the ECG signal from a particular person.

Whenever a new peak is detected, it must be categorized as a noise peak or a signal peak. If the peak level exceeds *THRESHOLD I1* during the first analysis of the signal, then it is a QRS peak. If searchback technique (explained in the next section) is used, then the signal peak should exceed *THRESHOLD I2* to be classified as a QRS peak. If the QRS complex is found using this second threshold level, then the peak value adjustment is twice as fast as usual:

$$SPKI = 0.25 \, PEAKI + 0.75 \, SPKI$$

The output of the final filtering stages, after the moving window integrator, must be detected for peaks. A peak detector algorithm finds peaks and a detection algorithm stores the maximum levels at this stage of the filtered signal since the last peak detection. A new peak is defined only when a level that is less than half the height of the peak level is reached. Figure 12.27 illustrates that this occurs halfway down the falling edge of the peak (Hamilton and Tompkins, 1986).

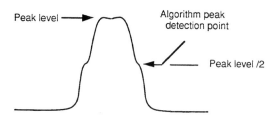

Figure 12.27 Output after the moving window integrator, with peak detection point.

12.5.6 Searchback technique

To implement the searchback technique, this algorithm maintains two RR-interval averages. One average, *RR AVERAGE1*, is that of the eight most recent heartbeats. The other average, *RR AVERAGE2*, is the average of the eight most recent beats which had *RR* intervals that fell within a certain range.

$$RR\ AVERAGE1 = 0.125\ (RR_{n-7} + RR_{n-6} + \ldots + RR_n)$$
$$RR\ AVERAGE2 = 0.125\ (RR'_{n-7} + RR'_{n-6} + \ldots + RR'_n)$$

The RR'_n values are the *RR* intervals that fell within the following limits:

$$RR\ LOW\ LIMIT = 92\% \times RR\ AVERAGE2$$
$$RR\ HIGH\ LIMIT = 116\% \times RR\ AVERAGE2$$

Whenever the QRS waveform is not detected for a certain interval, *RR MISSED LIMIT*, then the QRS is the peak between the established thresholds mentioned in the previous section that are applied during searchback.

$$RR\ MISSED\ LIMIT = 166\ \% \times RR\ AVERAGE2$$

The heart rate is said to be normal if each of the eight most recent RR intervals are between the limits established by *RR LOW LIMIT* and *RR HIGH LIMIT*.

12.5.7 Performance measurement

We tested the performance of the algorithm on the 24-hour annotated MIT/BIH database, which is composed of half-hour recordings of ECGs of 48 ambulatory patients (see references, MIT/BIH ECG database). This database, available on CD ROM, was developed by Massachusettsa Institute of Technology and Beth Israel Hospital. The total error in analyzing about 116,000 beats is 0.68 percent, corresponding to an average error rate of 33 beats per hour. In fact, much of the error comes from four particular half-hour tape segments (i.e., two hours of data from the total database).

Figure 12.28 shows the effect of excluding the four most problematic half-hour tapes from the overall results. Notice that the false-positive errors decrease much more than do the false negatives. This difference indicates that this algorithm is more likely to misclassify noise as a QRS complex than it is to miss a real event. Elimination of these four half-hour tape segments reduces the error rate below 10 beats per hour. Another available ECG database was developed by the American Heart Association (see references, AHA ECG database).

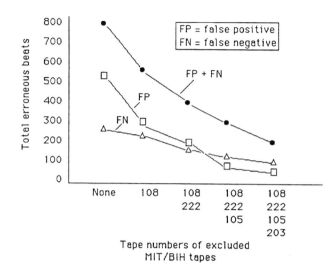

Figure 12.28 Performance of QRS detection software. The total error of the QRS detection algorithm can be substantially reduced by selectively eliminating problem tapes in the database.

12.6 LAB: REAL-TIME ECG PROCESSING ALGORITHM

This lab lets you "look inside" the inner workings of the algorithm QRS detection algorithm developed by Pan and Tompkins (1985) that is described in section 12.5. Load UW DigiScope, select `ad(V) Ops`, then `(Q)RS detect`.

12.6.1 QRS detector algorithm processing steps

Observe the output of each of the stages in the QRS detector. Sketch or print one cycle of the original ECG signal and the outputs of the low-pass, bandpass, derivative, squaring, and moving window integrator stages. Note the filter delay at each of these stages.

12.6.2 Effect of the value of the Q of a filter on QRS detection

Implement several two-pole recursive filters with 17-Hz center frequencies to observe the effects of different values of Q on the ECG, as in section 12.2.1. What value of r produces the most desirable response for detecting the QRS complex?

12.6.3 Integer filter processing of the ECG

Use (G) enwave to generate an ECG signal sampled at 100 Hz. Process this signal with a filter having the following difference equation.

$$y(nT) = 2y(nT - T) - 3y(nT - 2T) + 2y(nT - 3T) - y(nT - 4T)$$
$$+ x(nT) - 2x(nT - 12T) + x(nT - 24T)$$

Observe the output and note the duration of the ringing.

12.7 REFERENCES

AHA ECG database. Available from Emergency Care Research Institute, 5200 Butler Pike, Plymouth Meeting, PA 19462.

Ahlstrom, M. L. and Tompkins, W. J. 1983. Automated high-speed analysis of Holter tapes with microcomputers. *IEEE Trans. Biomed. Eng.*, **BME-30**: 651–57.

Ahlstrom, M. L. and Tompkins, W. J. 1985. Digital filters for real-time ECG signal processing using microprocessors. *IEEE Trans. Biomed. Eng.*, **BME-32**: 708–13.

Balda R. A., Diller, G., Deardorff, E., Doue, J., and Hsieh, P. 1977. The HP ECG analysis program. *Trends in Computer-Processed Electrocardiograms*. J. H. vanBemnel and J. L. Willems, (eds.) Amsterdam, The Netherlands: North Holland, 197–205.

Dobbs, S. E., Schmitt, N. M., Ozemek, H. S. 1984. QRS detection by template matching using real-time correlation on a microcomputer. *Journal of Clinical Engineering*, **9**: 197–212.

Friesen, G. M., Jannett, T. C., Jadallah, M. A., Yates, S. L., Quint, S. R., Nagle, H. T. 1990. A comparison of the noise sensitivity of nine QRS detection algorithms. *IEEE Trans. Biomed. Eng.*, **BME-37**: 85–97.

Furno, G. S. and Tompkins, W. J. 1982. QRS detection using automata theory in a battery-powered microprocessor system. *IEEE Frontiers of Engineering in Health Care*, **4**: 155–58.

Hamilton, P. S. and Tompkins, W. J. 1986. Quantitative investigation of QRS detection rules using the MIT/BIH arrhythmia database. *IEEE Trans. Biomed. Eng.* **BME-33**: 1157–65.

MIT/BIH ECG database. Available from: MIT-BIH Database Distribution, Massachusetts Institute of Technology, 77 Massachusetts Avenue, Room 20A-113, Cambridge, MA 02139.

Pan, J. and Tompkins, W. J. 1985. A real-time QRS detection algorithm. *IEEE Trans. Biomed. Eng.* **BME-32**: 230–36,.

Thakor, N. V., Webster, J. G., and Tompkins, W. J. 1983. Optimal QRS detector. *Medical and Biological Engineering*, 343–50.

Thakor, N. V., Webster, J. G., and Tompkins, W. J. 1984. Estimation of QRS complex power spectra for design of a QRS filter. *IEEE Trans. Biomed. Eng.*, **BME-31**: 702–05.

12.8 STUDY QUESTIONS

12.1   How can ectopic beats be detected using the automata approach to QRS detection?

12.2   How can QRS complexes in abnormal waveforms be detected using the crosscorrelation method?

12.3   In the moving window integrator of the algorithm in section 12.5, how should the width of the window be chosen? What are the effects of choosing a window width that is too large or too small?

12.4   In the QRS detection algorithm explained in section 12.5, how should the first threshold in each set of thresholds be changed so as to increase the detection sensitivity of irregular heart rates?

12.5   What are the effects of bandpass filter Q on the QRS-to-noise ratio in the ECG?

12.6   Design an algorithm that obtains the fiducial point on the ECG.

12.7   As an implementation exercise write a program using the C language, to detect QRS complexes in the ECG signal using any of the techniques described in this chapter.

12.8   Suggest a QRS detection algorithm, based on some of the techniques explained in this chapter or in other related literature, that can detect QRS complexes from the ECG in real time.

12.9   Experiments to determine the frequency characteristics of the average QRS complex have shown that the largest spectral energy of the QRS complex occurs at approximately what frequency?

12.10   A filter with the difference equation, $y(nT) = (y(nT - T))^2 + x(nT)$, is *best* described as what traditional filter type?

12.11   The center frequency of the optimal QRS bandpass filter is not at the location of the maximal spectral energy of the QRS complex. (a) What function is maximized for the optimal filter? (b) What is the center frequency of the optimal QRS filter for cardiotachometers? (c) If this filter has the proper center frequency and a $Q = 20$, will it work properly? If not, why not?

12.12   In addition to heart rate information, what QRS parameter is provided by the QRS detection algorithm that is based on the first and second derivatives?

12.13   The derivative algorithm used in a real-time QRS detector has the difference equation: $y(nT) = 2x(nT) + x(nT - T) - x(nT - 3T) - 2x(nT - 4T)$. (a) Draw its block diagram. (b) What is its output sequence in response to a *unit step* input? Draw the output waveform.

12.14   Write the equations for the amplitude and phase responses of the derivative algorithm used in a real-time QRS detector that has the transfer function

$$H(z) = \frac{-2z^{-2} - z^{-1} + z^{1} + 2z^{2}}{8}$$

12.15   A moving window integrator integrates over a window that is 30 samples wide and has an overall amplitude scale factor of 1/30. If a unit impulse (i.e., 1, 0, 0, 0, ...) is applied to the input of this integrator, what is the output sequence?

12.16   A moving window integrator is five samples wide and has a unity amplitude scale factor. A pacemaker pulse is described by the sequence: (1, 1, 1, 1, 0, 0, 0, 0, ...). Application of this pulse to the input of the moving window integrator will produce what output sequence?

12.17   The transfer function of a filter used in a real-time QRS detection algorithm is

$$H(z) = \frac{(1 - z^{-6})^2}{(1 - z^{-1})^2}$$

For a sample rate of 200 sps, this filter *eliminates* input signals of what frequencies?

13

ECG Analysis Systems

Willis J. Tompkins

This chapter covers the techniques for analysis and interpretation of the 12-lead ECG. Then it discusses ST-level analysis that is used in cardiac stress test systems. Finally, there is a summary of the hardware and software design of a portable ECG arrhythmia monitor.

13.1 ECG INTERPRETATION

Computer interpretation of the 12-lead ECG uses algorithms to determine whether a patient is normal or abnormal. It also provides written description of any abnormalities discovered.

13.1.1 Historical review of ECG interpretation by computer

ECG interpretation techniques were initially developed and used on mainframe computers in the early 1960s (Pordy et al., 1968). In those days, mainframe computers centrally located in computing centers performed the ECG analysis and interpretation. The ECGs were transmitted to the computer from remote hospital sites using a specially designed ECG acquisition cart that could be rolled to the patient's bedside. The cart had three ECG amplifiers, so three leads were acquired simultaneously and transmitted over the voice-grade telephone network using a three-channel analog FM modem. The interpretation program running in the mainframe computer consisted of several hundred thousand lines of FORTRAN code.

As technology evolved, minicomputers located within hospitals took over the role of the remote mainframes. The ECG acquisition carts began to include embedded microprocessors in order to facilitate ECG capture. Also, since the interpretation algorithms had increased failure rates if the ECG was noisy, the microprocessors increased the signal-to-noise ratio by performing digital signal preprocessing algorithms to remove baseline drift and to attenuate powerline interference.

Ultimately the ECG interpretation programs were incorporated within the bedside carts themselves, so that the complete process of acquisition, processing, and interpretation could be done at the patient's bedside without transmitting any data to a remote computer. This technology has now evolved into stand-alone microprocessor-based interpretive ECG machines that can be battery powered and small enough to fit in a briefcase.

The early ECG carts had three built-in ECG amplifiers and transmitted 2.5-second epochs of three simultaneous channels. In order to acquire all 12 leads, they sequenced through four groups of three leads each, requiring 10 seconds to send a complete record. Thus, the four acquired three-lead sets represented four different time segments of the patient's cardiac activity. Since a 2.5-second interval only includes two or three heartbeats, the early algorithms had difficulty in deducing abnormalities called arrhythmias in which several heartbeats may be involved in a rhythm disturbance. In order to improve arrhythmia analysis, three additional leads, typically the VCG leads, were recorded for a longer period of six seconds and added to the acquired data set (Bonner and Schwetman, 1968).

The modern microprocessor-based interpretive machines include eight ECG amplifiers so that they can simultaneously sample and store eight leads—I, II, and V1–V6. They then synthesize the four redundant leads—III, aVR, aVL, and aVF (see Chapter 2). These machines include enough memory to store all the leads for a 10-second interval at a clinical sampling rate of 500 sps.

13.1.2 Interpretation of the 12-lead ECG

ECG interpretation starts with feature extraction, which has two parts as shown in Figure 13.1. The goals of this process are (1) waveform recognition to identify the waves in the ECG including the P and T waves and the QRS complex, and (2) measurement to quantify a set of amplitudes and time durations that is to be used to drive the interpretation process. Since the computer cannot analyze the ECG waveform image directly like the human eye-brain system, we must provide a relevant set of numbers on which it can operate.

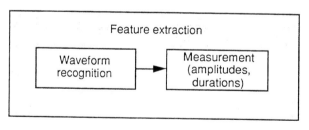

Figure 13.1 ECG feature extraction.

The first step in waveform recognition is to identify all the beats using a QRS detection algorithm (as in Chapter 12). Second, the similar beats in each channel are time-aligned and an average (or median) beat is produced for each of the 12 leads (see Chapter 9). These 12 average beats are analyzed to identify additional waves and other features of the ECG, and a set of measurements is then made and assembled into a matrix. These measurements are analyzed by subsequent processes.

Figure 13.2 is an ECG of a normal male patient acquired using an interpretive ECG machine. Although eight seconds of each lead are stored in the machine, only 2.5 seconds of each lead are printed on the paper copy as a summary. Figure 13.3 shows the internal measurement matrix that the ECG machine found for this patient. The amplitudes are listed in µV and the durations in ms. For example, in lead I, the R-wave amplitude (RA) is 1140 µV or 1.14 mV, and the R-wave duration (RD) is 71 ms.

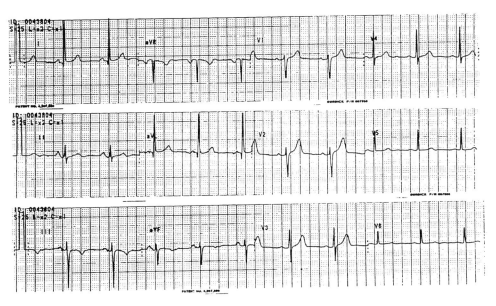

Figure 13.2 The 12-lead ECG of a normal male patient. Calibration pulses on the left side designate 1 mV. The recording speed is 25 mm/s. Each minor division is 1 mm, so the major divisions are 5 mm. Thus in lead I, the R-wave amplitude is about 1.1 mV and the time between beats is almost 1 s (i.e., heart rate is about 60 bpm). Recording was made on an Elite interpretive ECG machine (Siemens Burdick, Inc.).

	I	II	III	aVR	aVL	aVF	V1	V2	V3	V4	V5	V6
PA	70	127	77	-10	50	102	90	90	80	87	80	70
PPA	-10		-15	-90	-25		-25	-5		-3		
EOD	5			106	10							
OD	17			88	17							
OA	80			655	125							
RD	71	49	18		71	18	23	26	49	52	88	88
RA	1140	340	170		1055	130	250	430	1040	1560	1130	820
SD		39	70			9	82	57	31	32		
SA		110	970			80	1020	1190	760	390		
RPD						3						
RPA						30						
SPD						58						
SPA						430						
QRSA	681	105	-576	-399	823	-241	-783	-635	311	681	662	531
STJ	35	7	-28	-23	30	-10	42	70	55	12	10	20
STM	40	20	-20	-30	30		110	190	130	50	20	20
STE	50	30	-20	-40	35	5	180	320	230	90	30	20
TA	230	120	-130	-170	175	-25	480	860	700	330	150	110
TPA												
MTA	180	90	-110	-140	130	-30	300	540	470	240	120	90
MXFG												

Figure 13.3 Measurement matrix produced by an Elite interpretive ECG machine (Siemens Burdick, Inc.). Amplitudes are in μV and durations in ms.

There are two basic approaches for computerized interpretation of the ECG. The one used in modern commercial instrumentation is based on decision logic (Pordy et al., 1968; Macfarlane et al., 1971). A computer program mimics the human expert's decision process using a rule-based expert system. The second approach views ECG interpretation as a pattern classification problem and applies a multivariate statistical pattern recognition method to solve it (Klingeman and Pipberger, 1967).

Figure 13.4 shows the complete procedure for interpretation of the ECG. The feature extraction process produces a set of numbers called the measurement matrix. These numbers are the inputs to a decision logic or statistical process that drives an interpretation process which assigns words to describe the condition of the patient.

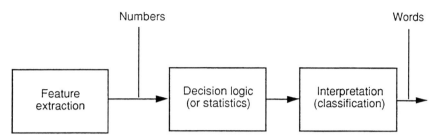

Figure 13.4 The steps in ECG interpretation.

The decision logic approach is based on a set of rules that operate on the measurement matrix derived from the ECG. The rules are assembled in a computer program as a large set of logical IF-THEN statements. For example, a typical decision rule may have the following format (Bonner and Schwetman, 1968):

```
Rule 0021:      IF
                (1) QRS ≥ .11 sec. on any two limb leads AND
                (2) Sd. ≥ .04 sec. on lead I or aVL  AND
                (3) terminal R present lead V1

                THEN
                (a) QRS .11 seconds;  AND
                (b) terminal QRS rightward and anterior; AND
                (c) incomplete right bundle branch block
```

The rules are usually developed based on knowledge from human experts. The pathway through the set of IF-THEN statements ultimately leads to one or more interpretive statements that are printed in the final report. Unfortunately, it is well known that a group of cardiologists typically interpret the same set of ECGs with less than 80 percent agreement. In fact, if the same ECGs are presented to one cardiologist at different times, the physician typically has less than 80 percent agreement with his/her previous readings. Thus, a decision logic program is only as good as the physician or group of physicians who participate in developing the knowledge base.

One advantage of the decision logic approach is that its results and the decision process can easily be followed by a human expert. However, since its decision rules are elicited indirectly from human experts rather than from the data, it is likely that such a system will never be improved enough to outperform human experts. Unlike human experts, the rule-based classifier is unable to make use of the waveforms directly. Thus, its capability is further limited to looking at numbers that are extracted from the waveforms that may include some measurement error. Also, with such an approach, it is very difficult to make minor adjustments to one or few rules so that it can be customized to a particular group of patients. Even slightly changing one rule may lead to modification of many different pathways through the logic statements.

For the multivariate statistical pattern recognition approach to ECG interpretation, each decision is made directly from the data; hence this approach is largely free from human influence. Decisions are made based on the probabilities of numbers in the ECG measurement matrix being within certain statistical ranges based on the known probabilities of these numbers for a large set of patients. Since this technique is dependent directly on the data and not on the knowledge of human experts, it is theoretically possible to develop an interpretive system that could perform better than the best physician.

However, unlike the decision logic approach, which can produce an explanation of how the decision is reached, there is no logic to follow in this approach, so it is not possible to present to a human expert how the algorithm made its final interpretation. This is the major reason that this technique has not been adopted in commercial instrumentation.

In clinical practice, physicians overread and correct computerized ECG interpretive reports. If similar waveforms are analyzed subsequently, the computer software makes the same diagnostic error over and over. Although it is desirable for an ECG interpretation system to "learn" from its mistakes, there is no current commercial system that improves its performance by analyzing its errors.

Figure 13.5 shows the final summary provided to the clinician by an interpretive ECG machine for the ECG of Figure 13.2. The machine has classified this patient as "Normal" with normal "Sinus rhythm."

```
I.D. : 0043804                          SINUS RHYTHM
50 Year-Old MALE                        NORMAL ECG
Med:


16:53 04/11/91 Loc:
Vent. Rate:                    61
PR interval                   180
QRS duration:                  92
QT/QTc:                    404/407
P-R-T axes:         52  20  -11
Limb: x2  Chest: x1
25 mm/s
```

Figure 13.5 Summary and interpretation (upper right corner) produced by an Elite interpretive ECG machine (Siemens Burdick, Inc.).

13.2 ST-SEGMENT ANALYZER

The ST-segment represents the period of the ECG just after depolarization, the QRS complex, and just before repolarization, the T wave. Changes in the ST-segment of the ECG may indicate that there is a deficiency in the blood supply to the heart muscle. Thus, it is important to be able to make measurements of the ST-segment. This section describes a microprocessor-based device for analyzing the ST-segment (Weisner et al., 1982).

Figure 13.6 shows an ECG with several features marked. The analysis begins by detecting the QRS waveform. Any efficient technique can be implemented to do this. The R wave peak is then established by searching the interval corresponding to 60 ms before and after the QRS detection mark, for a point of maximal value. The Q wave is the first inflection point prior to the R wave. This inflection point is recognized by a change in the sign of slope, zero slope, or a significant change in slope. The three-point difference derivative method is used to calculate the slope. If the ECG signal is noisy, a low-pass digital filter is applied to smooth the data before calculating the slope.

The isoelectric line of the ECG must be located and measured. This is done by searching between the P and Q waves for a 30-ms interval of near-zero slope. In order to determine the QRS duration, the S point is located as the first inflection point after the R wave using the same strategy as for the Q wave. Measurements of the QRS duration, R-peak magnitude relative to the isoelectric line, and the RR interval are then obtained.

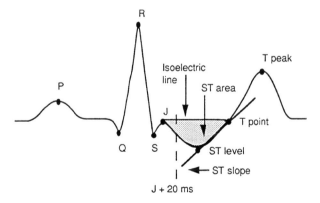

Figure 13.6 ECG measurements made by the ST-segment analyzer. The relevant points of the ECG, J, T, ST level etc. are indicated. The window in which the ST level is searched for is defined by J + 20 ms and the T point.

The J point is the first inflection point after the S point, or may be the S point itself in certain ECG waveforms. The onset of the T wave, defined as the T point, is found by first locating the T-wave peak which is the maximal absolute value, relative to the isoelectric line, between J + 80 ms and R + 400 ms. The onset of the T wave, the T point, is then found by looking for a 35-ms period on the R side of the T wave, which has values within one sample unit of each other. The T point is among the most difficult features to identify. If this point is not detected, it is assumed to be J + 120 ms.

Having identified various ECG features, ST-segment measurements are made using a windowed search method. Two boundaries, the J + 20 ms and the T point, define the window limits. The point of maximal depression or elevation in the window is then identified. ST-segment levels can be expressed as the absolute change relative to the isoelectric line.

In addition to the ST-segment level, several other parameters are calculated. The ST slope is defined as the amplitude difference between the ST-segment point and the T point divided by the corresponding time interval. The ST area is calculated by summing all sample values between the J and T points after subtracting the isoelectric-line value from each point. An ST index is calculated as the sum of the ST-segment level and one-tenth of the ST slope.

13.3 PORTABLE ARRHYTHMIA MONITOR

There is a great deal of interest these days in home monitoring of patients, particularly due to cost considerations. If the same diagnostic information can be obtained from an ambulatory patient as can be found in the hospital, it is clearly more cost effective to do the monitoring in the home. Technological evolution has led to a high-performance computing capacity that is manifested in such devices as compact, lap-sized versions of the personal computer. Such battery-powered systems provide us with the ability to do computational tasks in the home or elsewhere that were previously possible only with larger, nonportable, line-powered computers.

This increase in computing capability, with its concurrent decrease in size and power consumption, has led to the possibility of designing *intelligent* biomedical instrumentation that is small and light enough to be worn by an ambulatory patient (Tompkins, 1978; Webster, 1978; Webster et al., 1978; Tompkins et al., 1979; Tompkins, 1980; Tompkins, 1981a; Tompkins, 1981b; Tompkins, 1982; Tompkins, 1983). In addition, instrumentation can now be designed that was not contemplated previously, because microcomputer technology had not yet evolved far enough to support such applications (Sahakian et al., 1978; Tompkins et al., 1980; Weisner et al., 1982a; Weisner et al., 1982b; Chen and Tompkins, 1982; Tompkins et al., 1983). Portable instrumentation will ultimately perform monitoring functions previously done only within the confines of the hospital.

13.3.1 Holter recording

An initial goal is to replace the functions of the Holter tape recorder, the current device of choice for determining if an ambulatory patient has a potential cardiac problem. An optimal replacement device would be a microprocessor-based, portable arrhythmia monitor with processing algorithms similar to those now found in monitoring systems used in the cardiac care unit of today's hospital (Abenstein and Thakor, 1981; Thakor et al., 1984a).

Figure 13.7 shows the Holter approach, which is to record the ECG of a patient for a full day on magnetic tape. This recording and its subsequent return to a laboratory for playback and analysis restricts the timely reporting of suspected arrhythmias to the physician. The results of a Holter recording session are typically not known by the physician for several days.

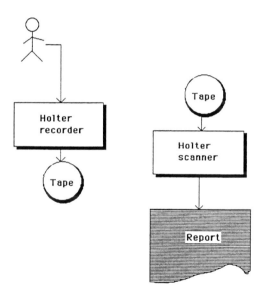

Figure 13.7 The contemporary Holter recording technique records the two-channel ECG for 24 hours on analog magnetic tape. A technician uses a Holter scanner to analyze the tape at 60 or 120 times real time and produces a final report.

What is called Holter monitoring is inappropriately named, as there is no monitoring in the normal context of the word. Monitoring typically provides continuous, real-time information about the state of the patient, as in intensive care monitoring.

The Holter recording approach to ambulatory ECG acquisition is a limited, low-technology technique. Although some improvements have been made in the small tape recorder itself, it is an electromechanical device with limited capability of improvement. Most of the technology changes have occurred in the central station Holter scanning equipment. However, there is still a great deal of manual intervention required to reduce the data captured to a useful form. This approach is labor intensive, and so it will remain for the foreseeable future.

13.3.2 Portable arrhythmia monitor hardware design

The *intelligent* portable arrhythmia monitor of Figure 13.8 will capture the ECG during suspected abnormal periods, and immediately send selected temporal epochs back to a central hospital site through the voice-grade telephone network.

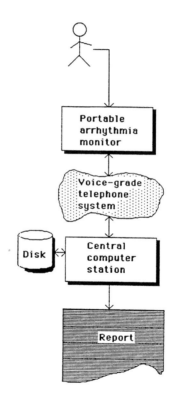

Figure 13.8 Portable arrhythmia monitor. This microprocessor-based device analyzes the ECG in real time and communicates captured data to a central host computer through the voice-grade telephone line.

This approach should provide a significant diagnostic edge to the cardiologist, who will be able to make judgments and institute therapeutic interventions in a much more timely fashion than is possible today. In addition, the clinician will be able to monitor the results of the therapy and modify it as needed, another factor not possible to do in a timely fashion with the tape recorder approach.

The hardware design of a portable arrhythmia monitor is quite straightforward, dependent only on the battery-operable, large-scale-integrated circuit components available in the marketplace. The primary semiconductor technology available for battery-operated designs is CMOS.

The hardware generations, however, evolve rapidly with the progressive improvements in semiconductor technology. For example, our initial portable arrhythmia monitor designs in 1977 were based on the COSMAC microprocessor (RCA CDP1802), a primitive central processing unit by today's standards. Modern commercial devices use the 80C86 microprocessor (Intel), a CMOS chip that is from the family of parts that make up the models of the IBM PC and compatibles.

Figure 13.9 shows a block diagram of one of our prototype monitors. In addition to the microprocessor, a portable arrhythmia monitor requires analog and digital support electronics. Analog amplifiers do the front-end ECG amplification and signal conditioning. An analog-to-digital (A/D) converter integrated circuit changes the analog ECG to the digital signal representation needed by the microprocessor.

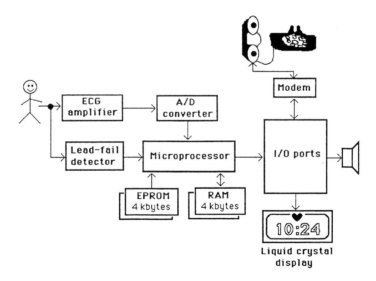

Figure 13.9 Block diagram of portable arrhythmia monitor.

ROM memory holds the program that directs the performance of all the functions of the instrument, and RAM memory stores the captured ECG signal. Input/output (I/O) ports interface audio and visual displays and switch interactions in the device. A modem circuit provides for communication with a remote computer so that captured ECG signals can be transmitted back to a central site (Thakor et al., 1982).

All these technologies are changing rapidly, so we must constantly consider improving the hardware design. Newer designs permit devices with (1) fewer components, (2) greater reliability, (3) smaller size and weight, (4) less power consumption, and (5) greater computational power (thereby permitting more complex signal processing and interpretive algorithms). Thus, there is a technological force driving us to continually improve the hardware design. In industry, an engineering compromise must come into play at some point, and the design must be frozen to produce a viable product. However, we are in a university environment, and so can afford to indulge ourselves by continual iteration of the design.

13.3.3 Portable arrhythmia monitor software design

Unfortunately, the frequent hardware changes lead to equally frequent software redesign. The software for an instrument is very hardware dependent. Each new microprocessor has its own unique machine language. Thus, we end up rewriting the same programs for different processors, and thereby waste considerable programming time. This software problem has led us to explore higher-level languages that are transportable from one kind of microprocessor to another.

We have settled on the C language, conceived at Bell Laboratories, as the most nearly ideal language now available for this type of real-time instrumentation application. Its primary advantages are (1) a programming level low enough to achieve the requisite machine control, and (2) transportability from one type of microprocessor to another. Providing that we follow a few software design rules, a real-time program written in the C language for one type of microprocessor can be easily reconfigured to run on a different one. Although it is not perfect, the C language considerably reduces the programming time necessary to rewrite the software when changing microprocessors. We then use our programming time to concentrate on improving the algorithms.

For a portable arrhythmia monitor, the two major software design tasks are (1) QRS detection and (2) arrhythmia analysis (Abenstein, 1978; Mueller, 1978). The QRS detection must be nearly perfect, otherwise the arrhythmia analysis algorithms will be fooled too often by false reports of beats that are not really there (i.e., false positives) or lack of reporting of beats that are missed (i.e., false negatives). An additional software design task important in these devices with limited memory is a data reduction algorithm (Tompkins and Abenstein, 1979; Abenstein and Tompkins, 1982).

QRS detection algorithm

Included in the various techniques that are used to implement a QRS detector are linear digital filters, nonlinear transformations, decision processes, and template matching (Thakor, 1978; Thakor et al., 1980; Ahlstrom and Tompkins, 1981; Furno and Tompkins, 1982; Thakor et al., 1983; Thakor et al., 1984b; Tompkins and Pan, 1985). Typically two or more of these techniques are combined together in a detector algorithm.

The most common approach in contemporary commercial ECG instrumentation is based on template matching. A model of the normal QRS complex, called a template, is extracted from the ECG during a learning period on a particular patient. This template is compared with the subsequent incoming real-time ECG to look for a possible match, using a mathematical criterion for goodness of fit. A close enough match to the template represents a detected QRS complex. If a waveform comes along that does not match but is a suspected abnormal QRS complex, it is treated as a separate template, and future suspected QRS complexes are compared with it. We have elected not to use this technique for the detection process, since it requires considerable memory for saving the templates (depending on how many you use; at least one system permits 40) and significant computational power for matching the templates to the real-time signal.

Instead, we use an algorithm based entirely on digital filters. Section 12.5 summarizes the QRS detection algorithm that we have developed for this application.

Arrhythmia analysis

From the QRS detector, the QRS duration and the RR intervals are determined. The ECG signal is then classified based on the QRS duration and the RR interval. Figure 13.10 is a conceptual drawing of an arrhythmia analysis algorithm based on the two parameters, RR interval and QRS duration (Ahlstrom and Tompkins, 1983). In this two-parameter mapping, we establish a region called *normal* by permitting the algorithm to first learn on a set of eight QRS complexes defined by a clinician as having *normal* rhythm and morphology for the specific patient. This learning process establishes the initial center of the *normal* region in the two-dimensional mapping space.

Boundaries of all the other regions in the map, except for region "0", are computed as percentages of the location of the center of the *normal* region. Region "0" has fixed boundaries based on physiological limits. Any point mapped into region "0" is considered to be noise because it falls outside what we normally expect to be the physiological limits of the smallest possible RR interval or QRS duration.

An abnormality such as tachycardia causes clusters of beats to fall in region "1" which represents very short RR intervals. Bradycardia beats fall in region "6". Typically, abnormalities must be classified by considering sequences of beats. For example, a premature ventricular contraction with a full compensatory pause would be characterized by a short RR interval coupled with a long QRS duration, followed by a long RR interval coupled with a normal QRS duration. This would be manifested as a sequence of two points on the map, the first in region "3" and the second in region "5". Thus, arrhythmia analysis consists of analyzing the ways in which the beats fall onto the mapping space.

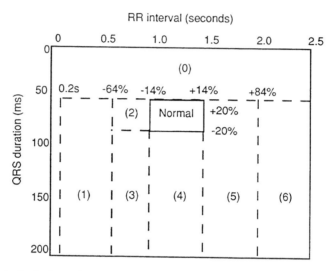

Figure 13.10 Arrhythmia analysis algorithm based on mapping the RR interval and QRS duration into two-dimensional space.

The center of the *normal* region is continuously updated, based on the average RR interval of the eight most-recent beats classified as normal. This approach permits the normal region to move in the two-dimensional space with normal changes in heart rate that occur with exercise and other physiological changes. The boundaries of other regions are modified beat-by-beat, since they are based on the location of the *normal* region. Thus, this algorithm adapts to normal changes in heart rate.

The classification of the waveforms can be made by noting the regions in which successive beats fall. Figure 13.11 lists some of the algorithms to detect different arrhythmias. The technique described is an efficient method for extracting RR interval and QRS duration information from an ECG signal. Based on the acquired information, different arrhythmias can be classified.

```
Normal:          If a beat falls in the normal box.
Asystole:        No R wave for more than 1.72 s; less than
                 35 beats/min
Dropped:         A long RR interval; beat falls in Region 6.
R-on-T:          A beat falls in Region 2.
Compensated      A beat in Region 3, followed by another in
 PVC:            Region 5.
Uncompensated    A beat in Region 3, followed by another
 PVC:            in the normal region.
Couplet:         Two consecutive beats in Region 3 followed
                 by a beat in the normal region, or
                 in Region 5.
Paroxysmal       If there are at least three consecutive
 Bradycardia:    points in Region 5
Tachycardia:     Average RR interval is less than
                 120 beats/min
Fusion:          A beat with a wide QRS duration;
                 falls in Region 4.
Escape:          A beat with a delayed QRS complex; falls in
                 Region  5.
Rejected:        A beat that has an RR interval of
                 200 ms or less, or QRS duration of
                 60 ms or less.
```

Figure 13.11 Classification of beats in the ECG signal based on QRS duration and RR interval.

13.3.4 The future of portable arrhythmia monitoring

One reason that is difficult to displace the old Holter technology with a modern high-technology approach is that the former is a full-disclosure technique that is inherently resistant to change. That is, even though the typical physician looks only at the final report and almost never looks at all the complete 24-hour ECG signal recorded, it is implicit that a skilled technician has analyzed the ECG. Also, the physician has the ultimate security blanket in that the data is there should it ever be necessary to go back through it again.

A second reason that the microprocessor-based, real-time approach has not affected the Holter market as yet is that the diagnostic algorithms are not fully perfected. The portable monitor must be able to do many of the tasks that are now being done in the coronary care unit. It must be able to detect reliably QRS complexes, perhaps better than hospital-based systems, since false judgments will cause unnecessary data to be stored in its limited memory. If problems occur, the portable device must perform self-diagnosis and make suggestions to the patient as to how to cure the problem. There is no technician in the ambulatory environment to correct problems when things go wrong.

Before physicians will rush to accept such a device, it must be clearly proven in clinical trials that it can capture the important clinical information at least as well as and at lower cost than the current Holter recording approach. Such a clinical demonstration is indeed difficult since there is no golden standard for proving performance. The MIT/BIH and AHA ECG databases serve only as preliminary evaluators.

Our current device does not attempt to make an interpretation of the patient's ECG. Instead it is designed to separate out suspicious waveforms from those that are considered normal for an individual patient. It then transmits the suspicious data by telephone to a clinician for a diagnostic judgment. Thus, it is a type of screening device. This approach as a first design step is simpler than a comprehensive ICU approach where critical clinical decisions must be made rapidly.

No portable arrhythmia monitor (or real-time Holter monitor as it is sometimes called) has yet been successful in the medical instrumentation marketplace. Part of the reason for this is that real-time QRS detection and arrhythmia analysis algorithms are not yet good enough for this application. However, there is steady progress in improvement of these algorithms, as microprocessors are being used more and more in bedside monitoring systems.

There is no doubt that Holter recording will be displaced, at some time, by microprocessor-based portable monitors. When that time comes, it will lead to lower diagnostic costs, greater device reliability, better clinical research capabilities, and continually evolving performance. The utility of these devices will evolve with the technology, just as the early four-function calculator has evolved into the lap-sized, high-performance portable computer of today.

13.4 REFERENCES

Abenstein, J. P. 1978. Algorithms for real-time ambulatory ECG monitoring. *Biomed. Sci. Instrum.*, **14**: 73–79.

Abenstein, J. P. and Thakor, N. V. 1981. A detailed design example—ambulatory ECG monitoring. In Tompkins, W. J. and Webster, J. G. (eds) *Design of Microcomputer-Based Medical Instrumentation.* Englewood Cliffs, NJ: Prentice Hall.

Abenstein, J. P. and Tompkins, W. J. 1982. A new data reduction algorithm for real-time ECG analysis. *IEEE Trans Biomed Eng.,* **BME-29**: 43–48.

Ahlstrom, M. L. and Tompkins, W. J. 1981. An inexpensive microprocessor system for high speed QRS width detection. *Proc. 1st Annual IEEE Compmed. Conf.,* 81–83.

Ahlstrom, M. L. and Tompkins, W. J. 1983. Automated high-speed analysis of Holter tapes with microcomputers. *IEEE Trans. Biomed. Eng.,* **BME-30**: 651–57.

Bonner, R. E. and Schwetman, H. D. 1968. Computer diagnosis of the electrocardiogram II. *Comp. Biomed. Res.,* **1**: 366.

Furno, G. S. and Tompkins, W. J. 1982. QRS detection using automata theory in a battery-powered microprocessor system. *IEEE Frontiers of Eng. in Health Care,* **4**: 155–58.

Klingeman, J., and Pipberger, H. V. 1967. Computer classification of electrocardiograms. *Comp. Biomed. Res.,* **1**: 1.

Macfarlane, P. W., Lorimer, A. R., and Lowrie, T. D. V. 1971. 3 and 12 lead electrocardiogram interpretation by computer. A comparison in 1093 patients. *Br. Heart J.,* **33**: 226.

Mueller, W. C. 1978. Arrhythmia detection program for an ambulatory ECG monitor. *Biomed. Sci. Instrum.,* **14**: 81–85.

Pan, J. and Tompkins, W. J. 1985. A real-time QRS detection algorithm. *IEEE Trans. Biomed. Eng.,* **BME-32**(3): 230–36.

Pordy, L., Jaffe,H., Chesky, K. et al. 1968. Computer diagnosis of electrocardiograms, IV, a computer program for contour analysis with clinical results of rhythm and contour interpretation. *Comp. Biomed. Res.*, 1: 408–33.

Thakor, N. V. 1978. Reliable R-wave detection from ambulatory subjects. *Biomed. Sci. Instrum.*, 14: 67–72.

Thakor, N. V., Webster, J. G., and Tompkins, W. J. 1980. Optimal QRS filter. *Proc. IEEE Conf. on Frontiers of Eng. in Health Care*, 2: 190–95.

Thakor, N. V., Webster, J. G., and Tompkins, W. J. 1982. A battery-powered digital modem for telephone transmission of ECG data. *IEEE Trans. Biomed. Eng.*, **BME-29**: 355–59.

Thakor, N. V., Webster, J. G., and Tompkins, W. J. 1983. Optimal QRS detector. *Med. & Biol. Eng. & Comput.*, 21: 343–50.

Thakor, N. V., Webster, J. G., and Tompkins, W. J. 1984a. Design, implementation, and evaluation of a microcomputer-based portable arrhythmia monitor. *Med. & Biol. Eng. & Comput.*, 22: 151–59.

Thakor, N. V., Webster, J. G., and Tompkins, W. J. 1984b. Estimation of QRS complex power spectra for design of a QRS filter. *IEEE Trans. Biomed. Eng.* **BME-31**: 702–706.

Tompkins, W. J. 1978. A portable microcomputer-based system for biomedical applications. *Biomed. Sci. Instrum.*, 14: 61–66.

Tompkins, W. J. 1980. Modular design of microcomputer-based medical instruments. *Med. Instrum.*, 14: 315–18.

Tompkins, W. J. 1981a. Portable microcomputer-based instrumentation. In Eden, M. S. and Eden, M. (eds) *Microcomputers in Patient Care*, Park Ridge, NJ: Noyes Medical Publications, 174–81.

Tompkins, W. J. 1981b. Role of microprocessors in ambulatory monitoring. *Proc. AAMI*, 99.

Tompkins, W. J. 1982. Trends in ambulatory electrocardiography. *IEEE Frontiers of Eng. in Health Care*, 4: 201–04.

Tompkins, W. J. 1983. Arrhythmia detection and capture from ambulatory outpatients using microprocessors. *Proc. AAMI*, 122.

Tompkins, W. J. and Abenstein, J. P. 1979. CORTES—A data reduction algorithm for electrocardiography. *Proc. AAMI*, 277.

Tompkins, W. J., Tompkins, B. M., and Weisner, S. J. 1983. Microprocessor-based device for real-time ECG processing in the operating room. *Proc. AAMI*, 122.

Tompkins, W. J., Webster, J. G., Sahakian, A. V., Thakor, N. V., and Mueller, W. C. 1979. Long-term, portable ECG arrhythmia monitoring. *Proc. AAMI*, 278.

Webster, J. G. 1978. An intelligent monitor for ambulatory ECGs. *Biomed. Sci. Instrum.*, 14: 55–60.

Webster, J. G., Tompkins, W. J., Thakor, N. V., Abenstein, J. P., and Mueller, W. C. 1978. A portable, microcomputer-based ECG arrhythmia monitor. *Proc. 31st ACEMB*, 60.

Weisner, S. J., Tompkins, W. J., and Tompkins, B. M. 1982a. A compact, microprocessor-based ECG ST-segment monitor for the operating room. *IEEE Trans. Biomed. Eng.*, **BME-29**: 642–49.

Weisner, S. J., Tompkins, W. J., and Tompkins, B. M. 1982b. Microprocessor-based, portable anesthesiology ST-segment analyzer. *Proc. Northeast Bioeng. Conf.*, 222–26.

13.5 STUDY QUESTIONS

13.1 In the modern version of the portable arrhythmia monitor, the arrhythmia analysis is based on mapping what two variables into two-dimensional space?

13.2 Current real-time QRS detection algorithms developed at the UW can correctly detect approximately what percentage of the QRS complexes in a standard 24-hour database?

13.3 In arrhythmia analysis, the RR interval and QRS duration for each beat are mapped into a two-dimensional space. How is the location of the center of the box marked *Normal* established?

13.4 Which of the following best describe the portable arrhythmia monitor developed at UW: (a) is a distributed processing approach, (b) selects important signals and stores them on magnetic tape for subsequent playback to a central computer over the telephone, (c) stores RR intervals and QRS durations in its memory so that a 24-hour trend plot can be made for these variables, (d) uses ST-segment levels as part of the arrhythmia analysis algorithm, (e) saves 30 16-second ECG segments in its memory, (f) transmits over the telephone using a separate modem that fits in a shirt pocket, (g) currently uses an CMOS 8088 microprocessor but will be updated soon, (h) always stores the ECG segment that preceded an alarm, (i) is being designed as a replacement for a Holter recorder, (j) uses the new medical satellite network to send its data to the central computer by telemetry, (k) has a built-in accelerometer for monitoring the patient's activity level, (l) includes 256 kbytes of RAM to store ECG signals, (m) uses two features extracted from the ECG in the arrhythmia analysis, (n) does near-optimal QRS detection so it will be produced commercially by a company early next year, (o) saves all the sampled two-channel ECG data for 24 hours, (p) stores the single ECG segment that caused an alarm, (q) analyzes the 12-lead ECG.

13.5 Which of the following best describe the portable arrhythmia monitor developed by Hewlett Packard: (a) a CMOS Z80 is the microprocessor, (b) the signal is transmitted to the central PC over the telephone, (c) sampled ECG waveforms are saved in RAM memory, (d) a 12-lead ECG is analyzed.

13.6 Describe the QRS detection technique that is used most in high-performance commercial arrhythmia monitors such as in the intensive care unit.

13.7 Explain how you would approach the problem of writing software to do 12-lead ECG interpretation by computer so that it would be commercially accepted.

13.8 What are some other techniques of measuring the ST-segment level? Give any advantages or disadvantages as compared to the windowed search method.

14

VLSI in Digital Signal Processing

David J. Beebe

The hardware used to implement the digital techniques discussed previously is the main focus of this chapter. We will first discuss digital signal processors (DSPs) and the functions they must perform. Next, two commercially available DSPs are described. Current high-performance VLSI architectures for signal processing are introduced including parallel processing, bit-serial processing, systolic arrays, and wavefront arrays. A portable ECG and a digital hearing aid are used as examples of VLSI applications in medicine. Finally, the emerging integration of VLSI and biomedical sensors is briefly discussed.

14.1 DIGITAL SIGNAL PROCESSORS

Until about 25 years ago, most signal processing was performed using specialized analog processors. As digital systems became available and digital processing algorithms could be implemented, the digital processing of signals became more widespread. Initially, digital signal processing was performed on general-purpose microprocessors such as the Intel 8088. While this certainly allowed for more sophisticated signal analysis, it was quite slow and was not useful for real-time applications. A more specialized design was needed.

14.1.1 Processor requirements and elements

Digital signal processors are really just specialized microprocessors. Microprocessors are typically built to be used for a wide range of general-purpose applications. In addition, microprocessors normally run large blocks of software, such as operating systems, and usually are not used for real-time computation.

A digital signal processor, on the other hand, is designed to perform a fairly limited number of functions, but at very high speeds. The digital signal processor must be capable of performing the computations necessary to carry out the techniques described in previous chapters. These include transformation to the frequency domain, averaging, and a variety of filtering techniques. In order to perform these operations, a typical digital signal processor would include the following elements:

1. Control processor
2. Arithmetic processor
3. Data memory
4. Timing control
5. Systems

In early digital signal processing systems, the implementation of the elements shown schematically in Figure 14.1 involved many chips or ICs (integrated circuits). Today all the elements can be realized on a single VLSI chip.

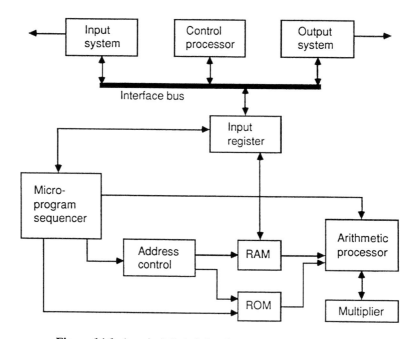

Figure 14.1 A typical digital signal processor modular design.

14.1.2 Single-chip digital signal processors

Some of the earliest attempts to incorporate the elements shown in Figure 14.1 on a single chip took place in the late 1970s. By the early 1980s, several companies including Bell Labs, Texas Instruments, NEC, and Intel had commercially available single-chip digital signal processors. In the last few years, the use of VLSI have made DSPs easier to use and more affordable.

Comparing the performance of DSPs is not always a straightforward procedure. While MIPS (million instructions per second) or MFlops (million floating-point operations per second) are often used when comparing microprocessor speed, this is not well suited to DSPs. A common benchmark for comparing the performance of DSPs is the multiply and accumulate (MAC) time. The MAC time generally reflects the maximum rate at which instructions involving both multiplication and accumulation can be issued. More meaningful benchmarks would be computations such as FFTs and digital filters. However, comparing these is tricky, because the benchmarks are not always completely described and often do not match those of the competition (Lee, 1988). Figure 14.2 shows the MAC times for some common DSPs.

The following sections describe two DSPs. First we discuss the TMS320 family made by Texas Instruments. This has been the most widely used DSP family. Second we describe the DSP56001 by Motorola. For a rough comparison of speed, consider that a Motorola 68000 microprocessor can handle 270,000 multiplications per second, while the DSP56001 is capable of 10,000,000 multiplications per second (Mo, 1991). That is an increase in speed of 37 times.

Company	Part	Date	MAC (ns)	Bits in mult.
AT&T	DSP1	1979	800	16
Texas Inst.	TMS32010	1982	390	16
Fujitsu	MB8764	1983	100	16
NEC	µPD77220	1986	100	24
Motorola	DSP56001	1987	74	24
AT&T	DSP16A	1988	33	16
Texas Inst.	TMS320C30	1988	60	24
Motorola	DSP96001	1989	75	32

Figure 14.2 Comparison of MAC times for several popular DSPs (adapted from Lee, 1988).

TMS320

TMS320 refers to a family of microprocessors introduced by Texas Instruments in 1982 and designed for application in real-time digital signal processing. The major feature of the TMS320 is a large on-chip array multiplier. In most general-purpose microprocessors, the multiplication is done in microcode. While this provides great versatility, it makes for long MAC times. An on-chip hardware multiplier greatly reduces the MAC times. On a typical microprocessor, approximately 10 percent of the chip area performs arithmetic functions, while on the TMS320, 35 percent of the chip area is dedicated to these functions. This large computing area is consistent with the numerically intensive nature of digital signal processing algorithms. The arithmetic and logic unit operates with 16/32-bit logic and the parallel hardware multiplication performs 16×16-bit two's complement multiplication in 200 ns. This high multiplication rate supports high-speed FFT computations. Programming is done in assembly language. The maximum clock rate is 20 MHz (Quarmby, 1985; Yuen et al., 1989).

DSP56001

The DSP56001 was introduced in 1987 and has quickly gained widespread use in audio equipment, scientific instrumentation, and other applications. The chip is capable of 10 million multiplications, 10 million additions, 20 million data movements, and 10 million loop operations per second. To achieve these high speeds, the chip has a highly parallel architecture with a hardware array multiplier, two ALUs (arithmetic logic units), two independent on-chip memory spaces, and an on-chip program memory. Most important, the DSP56001 utilizes a parallel and pipelined architecture in which several independent units operate simultaneously (Mo, 1991). The Motorola DSP is also unique in its ability to perform a MAC in just one cycle, with the result available by the next cycle. Its features include 512 words of on-chip program RAM, 24-bit data paths providing a dynamic range of 144 dB, a data ALU, address arithmetic units and program controller operating in parallel, and a MAC time of 74 ns (Lee, 1988; Motorola, 1988).

14.2 HIGH-PERFORMANCE VLSI SIGNAL PROCESSING

Only recently has it become feasible to perform digital signal processing in real time. This is because the implementation of digital processing techniques requires high levels of computational throughput, particularly for real-time applications. This demand combined with the continually increasing levels of performance of VLSI has led to the development of VLSI digital signal processors such as the TMS320 and the DSP56001 discussed above. The trend in DSP design is toward more algorithm-based architectures. In other words, the ease with which VLSI design can be done today leads the designer to more specialized architectures.

As discussed previously, the most useful digital signal processing techniques include FFT computing, FIR and IIR digital filters. The implementation of these techniques requires only three types of operations. The required operations are storage, multiplication, and addition. The small number of operations required suggest the use of a repetitive modular architecture. This is indeed the case. The limited number of different operations required and the way in which VLSI technology is fabricated have led to several VLSI-oriented special-purpose architectures for digital signal processing applications. These architectures use multiprocessing and parallel processing, array processors, reduced instruction set computer (RISC), and pipelining to achieve very high-speed processing rates.

The terminology used to describe and classify VLSI architectures is by no means standard. Various attempts have been made to establish a useful taxonomy, including Flynn's terminology based on instruction and data streams (Flynn, 1966). In the following discussion, we have tried to use the most commonly used terms. However, be aware that the literature is peppered with various terminology used to describe the similar architectures.

Parallel processing or multiprocessing uses multiple processors that cooperate to solve problems through concurrent execution. Pipelining is just an extension of multiprocessing that optimizes resource utilization and takes advantage of dependencies among computations (Fortes and Wah, 1987). Array processor typically refers to a two-dimensional array of processors that the data flows through. Pipelining is used to route the data through the array in the most efficient manner.

14.2.1 Parallel processing

The application of parallel processing to signal processing generally consist of two types (Yuen *et al.*, 1989). Figure 14.3 illustrates these architectures.

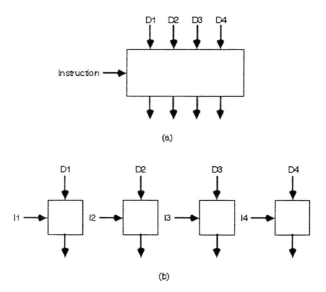

Figure 14.3 Classification of parallel processor architectures. (a) Single Instruction Multiple Data (SIMD), (b) Multiple Instruction Multiple Data (MIMD).

1. The single instruction multiple data (SIMD) operation in which the computational field consists of a number of data elements acted upon by a single operational instruction.

2. The multiple instruction multiple data (MIMD) operation in which a number of instruction streams act on multiple data elements. This method is used in image processing work.

Parallelism is achieved at the mathematical level by applying the residue number system (RNS) in architectural form (Szabo and Tanaka, 1967). In this system, the computational field is decomposed into a set of independent subfields. Computations in these subfields are performed in parallel in a SIMD-like structure. RNS provides for high-speed mathematical operations since addition and subtraction have no interdigit carries or borrows and multiplication does not require the generation of partial products.

14.2.2 Bit-serial processing

In bit-serial processors, operations are performed on only 1 bit in each word at a time. Bit-serial organization is usually applied to many words at once. Hence, the term bit-slice or bit-column arose. Figure 14.4 shows a simple bit-slice scheme.

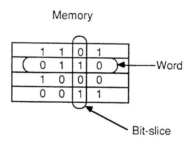

Figure 14.4 A simple bit-slice organization. The bit-slice of data is operated on, so one bit from each word is operated on simultaneously.

In bit-serial architecture, the digital signals are transmitted bit sequentially along single wires rather than simultaneously on a parallel bus (Denyer and Renshaw, 1985). This approach has several advantages over the parallel approach. First, communications within and between VLSI chips is more efficient. This is an important point given the communication-dominated operations involved in signal processing. Second, bit-serial architecture leads to an efficient pipelining structure at the bit level which leads to faster computations.

There are also several advantages to the bit-serial approach in terms of doing the actual VLSI chip layout. Bit-serial networks are easily routed on the chip since there is no need to make parallel connections to a bus. Also, since all signals enter and leave the chip via single pins, the number of input/output pins is reduced. Finally, bit-level pipelining distributes both memory and processing elements on the chip in a modular and regular fashion. This greatly facilitates ease of design, layout, and the application of silicon compilers (Yuen et al., 1989).

14.2.3 Systolic arrays

A systolic array takes a bit-serial architecture and applies pipelining principles in an array configuration. The array can be at the bit-level, at the word-level, or at both levels. The name systolic array arose from analogy with the pumped circulation of the bloodstream. In the systolic operation, the data coefficients and other information are systematically fed into the system with the results "pumped out" for additional processing. A high degree of parallelism is obtained by pipelining data through multiple processors, typically in a two-dimensional fashion. Once data enters the array, it is passed to any processor that needs it without any stores to memory. Figure 14.5 illustrates this data flow. The data flow is synchronized via a global clock and explicit timing delays (Duncan, 1990).

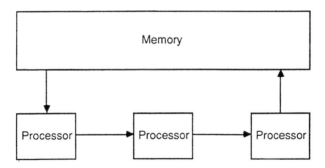

Figure 14.5 Systolic flow of data to and from memory.

Systolic arrays are particularly well suited to VLSI implementation. In order to take full advantage of the ever increasing density of VLSI, chip layouts must be simple, regular, and modular. Systolic arrays use simple processing elements and interconnection patterns that are replicated along one or two dimensions on the chip. In fact, most connections involve only nearest neighbor communication. The general architecture for a systolic array is shown in Figure 14.6. The chip resembles a grid in which each point is a processor and each line is the link between them.

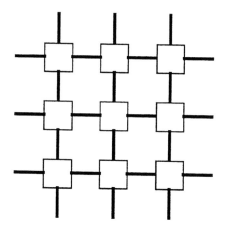

Figure 14.6 A simple systolic array configuration. Each box represents a processor or transputer and each line represents the link between the processors.

14.2.4 Wavefront arrays

One of the newest architectures is the wavefront array. Developed by Kung, the wavefront architecture is similar to the systolic array in that both are characterized by modular processors and regular, local interconnection networks. The difference is that in the wavefront array, the global clock and time delays are replaced by asynchronous handshaking. This eliminates problems of clock skew, fault tolerance, and peak power (Duncan, 1990).

   The details of these architectures is beyond the scope of this text; the intent here is to give the reader an overview of the current capabilities and applications for VLSI digital signal processing (see Kung et al., 1985 and Kung, 1988 for details of VLSI architectures).

14.3 VLSI APPLICATIONS IN MEDICINE

VLSI devices are currently used in a wide variety of medical products ranging from magnetic resonance imaging systems to conventional electrocardiographs to Holter monitors to instruments for analyzing blood.

14.3.1 Portable ECG

The portable ECG machine was made possible largely due to progress in VLSI design. Today's portable ECG machines, such as the Elite (Siemens Burdick, Inc.), are fully functional 12-lead ECG machines equal in every respect to larger machines with the exception of paper size and mass storage capacity. A quick comparison points out the dramatic results that can be obtained with VLSI. The portable ECG is 15 times smaller, 10 times lighter, and half as costly as the full-size machine. It consumes 90 percent less power and uses 60 percent fewer parts (Einspruch and Gold, 1989).

14.3.2 Digital hearing aid

A traditional analog hearing aid consists of the parallel connection of bandpass filters. To provide accurate compensation, a large number of filters are needed. Size, complexity, and cost all limit the number of filters. A hearing aid with a large number of adjustable components is also difficult to fit and adjust to the needs of each individual patient.

With the advent of general-purpose DSPs and application-specific integrated circuits (ASIC), it has become feasible to implement a hearing aid using digital technology. The digital implementation utilizes many of the ideas discussed previously including A/D converters, D/A converters, and digital filtering. The biggest advantage of the digital design is that the transfer function of the filter used to compensate for hearing loss is independent of the hardware. The compensation is performed in software and thus is extremely flexible and can be tailored to the individual. The digital hearing aid is more reliable and the fitting process can be totally automated (Mo, 1988).

14.4 VLSI SENSORS FOR BIOMEDICAL SIGNALS

A recent spinoff from standard microelectronics technology has been solid-state sensors and smart sensors. Integrated circuit processing techniques have been used for some time to fabricate a variety of sensing devices. Currently under development is a new generation of smart or VLSI sensors. These sensors combine the sensing element with signal processing circuitry on a single substrate. Figure 14.7 shows the elements of a generic VLSI sensor.

Integrated solid-state sensors or smart sensors contain four parts: (1) custom thin films for transduction, structural support or isolation, or encapsulation; (2) microstructures formed using micromachining techniques; (3) interface circuitry; and (4) signal processors (Einspruch and Gold, 1989). A sensor built by Najafi and Wise to measure the extracellular potentials generated within neurons illustrates the possibilities of combining VLSI with current sensor technologies.

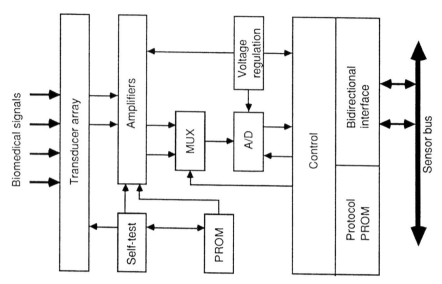

Figure 14.7 Block diagram of a generic VLSI sensor. The sensor is addressable, self-testing and provides a standard digital output (adapted from Einspruch and Gold, 1989).

Figure 14.8 shows the overall structure of the sensor. The electronics on the sensor includes functional circuits such as amplifiers, an analog multiplexer, a shift register, and a clock. Indeed most of the sampling functions discussed in Chapter 3 are integrated right onto the sensor. The sensor is 3–4 mm long, 200 μm wide and 30 μm thick.

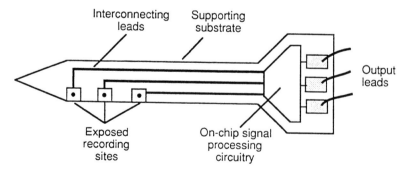

Figure 14.8 The overall structure of the smart sensor built by Najafi and Wise (adapted from Einspruch and Gold, 1989).

14.5 VLSI TOOLS

The most commonly used VLSI design tools in the academic environment are the Berkeley VLSI tools. The system allows the designer to lay out the design at the device level and simulate the design for correct operation. A typical design would begin with a functional block diagram. One next moves progressively lower and lower toward the actual device layout. The device level layout is done in MAGIC, a VLSI layout editor. MAGIC is analogous to a mechanical CAD program. The designer actually draws each and every transistor in the design just as it will be fabricated. Obviously this can be a painstaking task since a typical chip may contain several hundred thousand transistors. MAGIC contains many features that simplify the design and often portions of the layout are redundant. As portions of the design are completed, programs such as CRYSTAL (a VLSI timing analyzer) and ESIM (an event-driven switch level simulator) are used to test the logic and timing operation of that portion of the design. This is an important point! It is important to thoroughly test each portion of the layout as it is completed rather than waiting until the whole design is done. Continual testing will help ensure that the design will be functional when all the pieces are connected together in the final layout.

Once the entire layout is complete and it has been successfully simulated, it is fabricated. The MOSIS (MOS implementation system) fabrication facility at the University of Southern California is often used for low-volume experimental work. Finally, the device is tested and revised as necessary.

14.6 CHOICE OF CUSTOM, ASIC, OR OFF-THE-SHELF COMPONENTS

When approaching an instrument design, one must decide between simply using off-the-shelf chips, ASIC (application specific integrated circuits), or fully custom chips. In reality, the finished design will usually contain some combination of these approaches. Ten years ago, using off-the-shelf chips was the only option. However, in recent years it has become feasible for even small companies to design their own ASIC or fully custom chips via tools similar to those discussed above. The design choices are based on the needs of the particular project and the availability of suitable off-the-shelf chips. The design time and cost generally increase as one moves from a design containing only off-the-shelf chips to a fully custom design.

14.7 REFERENCES

Denyer, P. and Renshaw, D. 1985. *VLSI Signal Processing: A Bit-serial Approach.* Reading, MA: Addison-Wesley.

Duncan, R. 1990. A survey of parallel computer architectures. *Computer*, **23**(2): 5–16.

Einspruch, N. G. and Gold, R. D. 1989. *VLSI in Medicine*. San Diego: Academic Press.

Flynn, M. J. 1966. Very high speed computing systems. *Proc. IEEE*, **54**: 1901–09.

Fortes, J. A. B. and Wah, B. W. 1987. Systolic arrays—from concept to implementation. *Computer*, **20**(7) :12–17.

Kung, S. Y. 1988. *VLSI Array Processors*. Englewood Cliffs, NJ: Prentice Hall.

Kung, S. Y., Whitehouse, H. J., and Kailath, T. 1985. *VLSI and Modern Signal Processing*. Englewood Cliffs, NJ: Prentice Hall.

Lee, E. A. 1988. Programmable DSP architectures: Part 1. *IEEE ASSP Magazine*. 4–19.

Mo, F. 1991. Basic digital signal processing concepts for medical applications. *Biomedical Science and Technology*. **1**(1): 12–18.

Motorola. 1988. Technical Data #DSP56001/D, Motorola Literature Distribution, P.O. Box 20912, Phoenix, Arizona, 85036.

Quarmby, D. (ed.) 1985. *Signal Processor Chips*. Englewood Cliffs, NJ: Prentice Hall.

Szabo, N. S. and Tanaka, R. I. 1967. *Residue Arithmetic and Its Applications to Computer Technology*. New York: McGraw-Hill.

Yuen, C. K., Beauchamp, K.G., and Fraser, D. 1989. *Microprocessor Systems in Signal Processing*. San Diego: Academic Press.

14.8 STUDY QUESTIONS

14.1 Describe the difference between general-purpose microprocessors and DSPs.

14.2 Why are MAC times used as benchmarks for DSPs instead of the usual MIPS or MFlops?

14.3 Discuss why VLSI is well suited to the design of DSPs.

14.4 Define the following terms: (1) parallel processing, (2) pipelining, (3) array processors, (4) SIMD, (5) MIMD.

14.5 Describe the difference between systolic arrays and wavefront arrays. Which would operate more efficiently?

14.6 List at least four advantages of using VLSI-based designs for medical applications.

Appendix A

Configuring the PC for UW DigiScope

Danial J. Neebel

This appendix provides the steps necessary to set up the various hardware systems that are supported by UW DigiScope. DigiScope can interface to three different types of devices: (1) the Real Time Devices ADA2100 analog and digital I/O interface card that is installed internally in the IBM PC or compatible, (2) an external single board computer with serial RS232 communications (Motorola S68HC11EVBU Student Design Kit and MCM68HC11EVB Evaluation Kit), and (3) a virtual I/O device (data files). When you install UW DigiScope with the `INSTALL.EXE` program, you identify which physical devices are available in your system. This selection can be changed later if you add a conversion device.

Figure A.1 shows the three different devices and the capabilities of each. A sampling rate of approximately 500 samples per second (sps) can be achieved with a 12-MHz 80286-based IBM PC/AT compatible. The maximum sampling rate of a slower 8088-based IBM PC or compatible running at 4.77 MHz is approximately 250 sps. Thus, depending on your driving habits, mileage may vary.

Item	Internal device (ADA 2100)	External device (Motorola EVBU)	Virtual I/O device
Sampling rate (sps)	1 to 500	31 to 500	1 to 500
Analog input range (V)	−5 to +5	0 to +5	N/A
Number of analog inputs	8	4	20
Number of analog outputs	2	none	none
Analog output range (V)	−10 to +10	N/A	N/A
Number of digital inputs	4	8	none
Number of digital outputs	4	8	none

Figure A.1 Comparison of the three types of devices.

An important step in installing any of the three devices in your system is running the INSTALL program. INSTALL sets up all the software for DigiScope. If the Real Time Devices ADA2100 is installed in the system, you should place a call to DACINIT.COM in the AUTOEXEC.BAT. DACINIT.COM resets the ADA2100 timer so that the timer does not cause interrupts. The documentation with the ADA2100 explains why it is necessary to do this.

In the following three sections, we provide the steps necessary to get a PC system ready for Data Acquisition using DigiScope. First we cover configuring and installing the ADA2100. Next we present the steps to connect the Motorola EVBU and EVB to a PC using RS232 communications. Finally, a few pointers are presented for setting up a good system to perform Virtual Data Acquisition. Before discussing any of these hardware setups, it is important to know how to organize the work area correctly to do the job right and safely. Read through the information in the following box.

Warning: Preparing a Work Area

Safety first. Unplug all devices that you are working on. Never remove the case of any device without making sure that the device is unplugged. This will help protect you as well as your equipment.

Installing and configuring your system for either the ADA2100 or the Motorola EVBU will require you to handle circuit boards containing delicate integrated circuits (ICs). Proper care should be taken to limit the likelihood of any device becoming damaged. Start with a clear work area. Items such as an antistatic mat and wristband can help limit static electricity. If you do not have these items then always make sure you have grounded yourself before touching any ICs or circuit boards. This will help reduce the amount of static electricity in your body.

Many of the ICs on the ADA2100 and the EVBU are CMOS technology. Even a small amount of static electricity can damage a CMOS part permanently. The part may not show any signs of damage but it will fail to operate properly.

A.1 INSTALLING THE REAL TIME DEVICES ADA2100 IN AN IBM PC

Installing the Real Time Devices ADA2100 interface card in an IBM PC is a three-step process. Each step must be taken with great care. The first step is to configure the ADA2100. Next is the physical installation of the ADA2100 into the bus of the IBM PC. The final step is to set up the CONFIG.WDS file. This is done automatically by the INSTALL.EXE program and can be changed by running ADINSTAL.EXE.

A.1.1 Configuration of the ADA2100

DigiScope makes some assumptions about how the ADA2100 is configured. Figure A.2 gives the jumper and switch settings for the device. It is also important to check the rest of the internal cards in the system to determine if the ADA2100 settings will conflict with devices already installed in your system. The settings for the ADA2100 are flexible enough that this should not be a problem. ADA2100 interface card

Jumper/Switch	Function	Required settings
P2	Base I/O address	As defined by CONFIG.WDS
S1	Analog input signal type	Pos 1, 2, 3 UP and Pos 4 down
P3	PIT I/O header connector	Chain the timers together; see text below
P5	PIT interrupt header	As defined by CONFIG.WDS
P6	EOC monitor header	PA7
P7	EOC interrupt header	EOC not connected to any IRQ
P9	A/D converter voltage	10 V
P10	D/A converter voltage	Both set to ±

Figure A.2 RTD ADA2100 jumper configurations. ADA2100 interface card

The Base I/O Address (P2) must be set to the value defined during installation of the software. The analog input signal type (S1) should be set so that there are eight single-ended channels with ± polarity. The PIT I/O Header (P3) connector should be set so that OUT0 is connected to CLK1, OUT1 is connected to CLK2, and OUT2 can be connected to CO2 or –CO2. The EG0, EG1, and EG2 lines should be connected to +5 V. The PIT interrupt header, P5, must be configured so that OUT2 is connected to the IRQ line defined during installation. The EOC monitor header, P6, should be set to PA7. The EOC interrupt header, P7, must be configured so that EOC is not connected to any interrupt. The A/D converter voltage range, P9, should be set to 10 V. Finally, the D/A converter output voltage range, P10, should have AOUT1 and AOUT2 both connected to ±.

A.1.2 Installing the ADA2100

In this section we do not give step-by-step instructions for removing the cover and installing a card. We assume that the reader is familiar enough with an IBM PC to be able to perform these tasks. If not, your system manual should give a good explanation of how to install an I/O card. In this section we give some indications of things to look out for. It is important to make sure that no other cards conflict with the settings on the ADA2100. The first of these is the Base I/O address. Make sure that no other card in the system is configured to use the addresses in the range Base I/O Address to Base I/O Address +0x17. If, for example, the Base I/O Address of the ADA2100 is 0x240, then there should be no other cards using addresses between 0x240 and 0x257. These addresses must be dedicated to the ADA2100. Also make sure no other card is using the IRQ line set on P5, the PIT Interrupt Header.

A.1.3 Installing ADA2100 power-up initialization

The AUTOEXEC.BAT file must be modified to execute DACINIT.COM. This can be done with almost any text editor. DACINIT.COM will set the 82C54 timer on the ADA2100 to a known state. The 82C54 timer does not automatically reset to a known state on power up. This makes it possible for the 82C54 to power up in a mode that will interrupt the system if the system does not disable interrupts. The system boot should disable this interrupt and set the interrupt vector to a "dummy" interrupt. If for some reason the system does not perform these operations, problems could occur. By running DACINIT.COM the 82C54 will not cause an interrupt. Also, DigiScope will disable the 8254 on exit. If DigiScope is not allowed to exit properly (Control Break is used), then the timer may still be running and causing interrupts to occur.

A.1.4 Connecting to the ADA2100

Figure A.3 shows the connection points for the inputs and outputs. The "–" signal pins are the ground lines. All the ground lines are the same on the ADA2100.

Signal	+ Signal pins	– Signal pins
Analog input channels 1 to 8	1 to 8	21 to 28
Analog output channel 1, 2	10, 11	30, 31
Digital inputs 0 to 3	36, 16, 35, 15	37
Digital outputs 0 to 3	34, 14, 33, 13	37

Figure A.3 Signal connections for the ADA2100.

Thus, if several channels share the same ground, only one ground lead need be connected to the ADA2100. The analog inputs are configured as eight single ended ± polarity channels. Do not exceed the input voltage range of ±5 V. Digital signals are to/from an 82C55 directly. Do not exceed the 0 to +5 V level of the 82C55. Also, a CMOS device like the 82C55 cannot drive a large amount of current. We recommend using a noninverting buffer such as the 74LS244 to protect the 82C55. See Appendix C of the ADA2100 User's Manual for more information on the inputs and outputs of the 82C55.

A.2 CONFIGURING THE MOTOROLA 68HC11EVBU

The Motorola EVBU Student Design Kit was chosen as an inexpensive data acquisition and control unit. This device was chosen because of its availability and low cost. At the time of this printing, Motorola is selling this kit for $68.11. This price is not likely to change. The EVBU is designed to emulate a Motorola 68HC11. The processor used in the EVBU has 8 channels of 8-bit A/D, an RS232 Serial Communications Interface, and 24 bits of digital I/O. As implemented in DigiScope, the EVBU has 4 channels of A/D (0 to +5 V). The capability exists for 8 digital inputs and 8 digital outputs (0 to +5 V).

Connecting the EVBU to a host computer is straightforward. All that is required is a cable and a +5-V dc power supply. The +5-V supply can be replaced by a battery. Instructions for using a battery to power the EVBU are included with the kit.

The EVBU does not come with a cable, but a good explanation of how to make a cable to connect the EVBU to the IBM PC is included with the kit. We offer a different cable design. Since our software does not require hardware handshaking, a cable with only three wires and some jumpers can be constructed. The jumpers are to ensure proper operation of PC serial communication cards which expect hardware handshaking. Figure A.4 shows cable designs for both 25-pin and 9-pin PC connectors.

A.2.1 Configuration of the EVBU

The initial configuration is shown in Figure A.5. Some minor changes will be made later after some setup of the program memory in the EVBU. All of these settings are defaults (or can be) except for J2 and J4. J2 need not be removed but a jumper is needed across J4 to put the 68HC11 into bootstrap mode. J2 is used by the BUFFALO monitor to determine if the monitor should be executed or if the processor should jump to the EEPROM at location 0xB600. Simply remove J2 and place it across J4.

Where the words installed or removed are used, there is a physical jumper that must be handled. Where the words shorted or opened are used, there is no physical jumper supplied but there may be a cut-trace short located on the printed circuit board solder side (bottom). It is not necessary to cut any traces on the circuit board. In fact, the only jumpers that need be changed are J2 and J4.

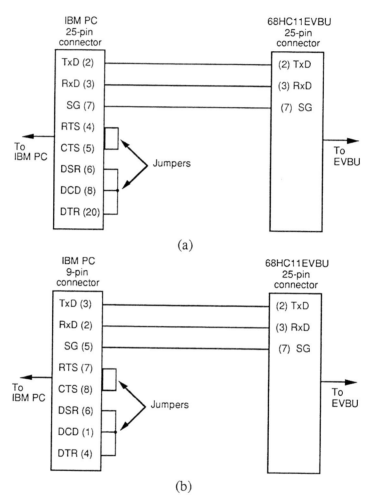

Figure A.4 Cable designs for connecting a PC to the 68HC11EVBU board. (a) 25-pin PC connector. (b) 9-pin PC connector.

Jumper	Function	Required settings
J1	Input power select header	Set as required for supply type
J2	Program execution select	Must be moved to J3
J3,J4	MCU mode select header	Must both be installed
J5,J6	MCU clock reconfiguration	Must both be open
J7	Trace enable header	Can be installed or removed
J8,J9	SCI reconfiguration	Both shorted
J10,11,12,13	SPI reconfiguration headers	Can be installed or removed
J14	Real time clock INT* header	Must be open
J15	TxD reconfiguration header	Must be shorted
J5	Terminal baud rate select	Across pins 11 and 12
J6	Host port Rx signal disable	Can be installed or removed

Figure A.5 Motorola EVBU jumper configurations

A.2.2 Connecting the IBM PC to the EVBU

You will need to provide a cable as described in the Motorola EVBU User's Manual. The Digiscope program checks the file CONFIG.WDS to determine the serial port used to communicate with the EVBU. The default is serial port configured as COM1. You should run the ADINSTAL.EXE program to create CONFIG.WDS if you have not done so already.

A.2.3 Installing EDAC 68HC11 program into the EVBU

The files EDAC.S19 and EDAC.ASM are on the disk that you received with this textbook. EDAC.ASM is the source code for the program that will reside on the EVBU and communicate with DigiScope. EDAC.S19 contains the hex code for EDAC.ASM. EDAC.S19 is in Motorola S-record format. A good description of the S-record file format can be found in Appendix A of the EVBU manual.

Included with the student project kit is a software development utility called pcBug11. pcBug11 can be used to develop software for the 68HC11. pcBug11 can be used to download programs from a PC to the 68HC11 RAM, EPROM or EEPROM using the 68HC11 bootstrap mode. A macro called LOAD.MCR for pcBug11 has been provided with DigiScope to allow simple programming of the 68HC11 EEPROM with EDAC.S19. After loading EDAC.S19 into the EEPROM of the 68HC11, it is necessary to use pcBug11 to start the EDAC program running on the 68HC11.

If the above is not workable, then the monitor may be programmed into the EPROM using a 12 V power supply and a 100 Ω resistor. With the monitor programmed into the EPROM, J2 may be configured such that the processor will start executing the code in EEPROM after reset. This means that the reset switch need only be pushed instead of rerunning pcBug11.

Important: Starting EDAC

Each time the EVBU is reset it is necessary to use
`pcBug11` to restart the `EDAC` program. The command to do
this is:

PCBUG11 -E port=N macro=go.

Where **N** is the number of the COM port that the EVBU
is connected to. **N** must be either 1, 2, 3, or 4. Also the
following files must be in the current directory:

PCBUG11.EXE, TALKE.XOO, TALKE.BOO, and **GO.MCR.**

All these files except for **GO.MCR** are included on the
`pcBug11` disk you received with the EVBU. **GO.MCR** is
included on your DigiScope disk.

A.2.4 Connecting signals to the EVBU

Figure A.6 shows the pin numbers for connecting analog and digital signals to the
Motorola EVBU. Note that analog input channels 1 to 4 are connected to port E
bits 4 through 7.

Signal	+ Signal pins	– Signal pins
Analog input channels 1 to 4	44, 46, 48, 50	1
Digital inputs 0 to 7	9-16	1
Digital outputs 0 to 7	42, 41, 40, 39, 38, 37, 36, 35	1

Figure A.6 Signal connections for the Motorola EVBU.

Although it is not absolutely necessary, we recommend to those using an EVBU
for data acquisition and control to add some protection to the inputs and outputs.
CMOS devices such as the 68HC11 are very sensitive to voltages outside the range
of 0 to +5 V dc and are not able to sink or source a large amount of current.

Protecting analog inputs

Figure A.7 gives an example of a simple protection circuit for analog inputs. To
keep the input voltage from rising above +5 V or dropping below 0 V, we have

suggested clamping diodes. Point A will not go below –0.7 V or above +5.7 V. This is enough to protect the 68HC11. Between the clamping diodes and the input on the IC, a current limiting resistor is used. A value of 10 kΩ will limit the current without changing the input impedance too much. Finally, a 0.1 μF ceramic capacitor is placed from the input to ground. This capacitor will help dampen the effects of any high-frequency noise that may be present.

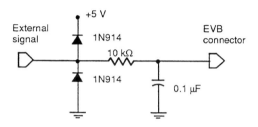

Figure A.7 Example of a simple protection circuit for a Motorola EVBU analog input.

Protecting digital inputs and outputs

Digital inputs and outputs are much easier to protect. A noninverting buffer can be placed between the outside world and the microcontroller. If an input or output is subjected to a damaging voltage, the buffer will be damaged and not the microcontroller. Also, a buffer made of Low Power Schottky TTL (LSTTL) technology is more resistant to out-of-range voltages. A 74LS244 works very well as a buffer. There are 8 buffers in a single package. Only one 74LS244 is required for each of the input and output banks.

A.3 CONFIGURING THE MOTOROLA 68HC11EVB

Setting up the Motorola EVB Evaluation Kit is much the same as setting up the system with the EVBU Student Project Kit. The EVB is also designed to emulate a Motorola 68HC11.

There are different jumpers to configure on the EVB since this is a different circuit board. The software setup for the EVB is similar to the software setup for the EVBU. The major difference is that a monitor program is already programmed into an external EPROM on the EVB. There is no need to program the internal EPROM of the 68HC11. The monitor is then used to program the EEPROM.

A.3.1 Setting up the EVB

There are two steps to configuring the EVB. The first is setting jumpers and building and connecting the RS232 cable. The second step is to connect the power supplies to the EVB. The EVB requires +5, +12, and -12 V supplies. The jumpers on the EVB should be configured as shown in Figure A.8. All of these settings are defaults (or can be). J4 will need to be changed to EEPROM after EEPROM has been programmed with EDAC.S19. This jumper is used by the BUFFALO monitor to determine if the monitor should be executed or if the processor should jump to the EEPROM at location 0xB600. Initially the jumper should be in position TBD to indicate that the program in the EPROM should be executed.

Jumper	Function	Required settings
J1	Reset select header	Can be installed or removed
J2	Clock select header	Across pins 2 and 3
J3	RAM select header	Can be installed or removed
J4	Program execution select	Must set to EPROM initially
J5	Terminal baud rate select	Across pins 11 and 12
J6	Host port Rx signal disable	Can be installed or removed

Figure A.8 Motorola EVB jumper configurations.

A.3.2 Installing EDAC 68HC11 program into the EVB

The files EDAC.S19 and EDAC.ASM are on the disk you received with this textbook. EDAC.ASM is the source code for the program that will reside on the EVBU and communicate with Digiscope. EDAC.S19 contains the hex code for EDAC.ASM. EDAC.S19 is in Motorola S-record format, A good description of the S-record file format can be found in Appendix A of the EVBU manual.

Included with the student project kit is a software development utility called pcBug11. pcBug11 can be used to develop software for the 68HC11. pcBug11 can be used to download programs from a PC to the 68HC11 RAM or EPROM using the 68HC11 bootstrap mode. A macro called LOAD.MCR for pcBug11 has been provided with Digiscope to allow simple programming of the 68HC11 EEPROM with EDAC.S19. After loading EDAC.S19 into the EEPROM of the 68HC11 it is necessary to use pcbug11 to start the EDAC program running on the 68HC11.

A.3.3 Connecting the IBM PC to the EVB

You will need to provide a cable as described in the Motorola EVB User's Manual. The DigiScope program checks the file CONFIG.WDS to determine the serial port used to communicate with the EVB. The default is serial port configured as COM1. You should run the ADINSTAL.EXE program to create CONFIG.WDS if you have not done so already.

A.3.4 Connecting signals to the EVB

See section A.2.4 for a description of how to connect signals to the EVB. The same basic principles apply for protecting analog inputs and digital output and digital inputs. Figure A.9 shows the pin numbers for connecting analog and digital signals to the Motorola EVB.

Signal	+ Signal pins	− Signal pins
Analog input channels 1 to 8	43, 45, 47, 49, 44, 46, 48, 50	1
Digital inputs 0 to 7	9-16	1
Digital outputs 0 to 7	42, 41, 40, 39, 38, 37, 36, 35	1

Figure A.9 Signal connections for the Motorola EVB.

A.4 VIRTUAL INPUT/OUTPUT DEVICE (DATA FILES)

Although it may seem a bit strange to talk about hardware configuration for a virtual device, this section gives you some indication of how a PC should be set up to run a virtual I/O device. The minimum requirements are a PC with 640 kbytes of memory, VGA or Hercules monochrome graphics, and a single floppy drive. For best performance, we strongly recommend using the program with a hard disk drive. If DigiScope must read from a floppy disk every time it reads a sample point then "real-time data acquisition" will not look very real.

A.5 PUTTING A HEADER ON A BINARY FILE: ADDHEAD

Files created by DigiScope have a unique structure. You can view information in a file header with the UW DigiScope `stat(U)s` command. A special program is also provided for creating file headers.

If you have a data file that contains 16-bit integers in strictly binary format (e.g., C-language 16-bit integers), you can use `ADDHEAD` to put a header on the file. `ADDHEAD` will prompt the user for the information to be placed in the header of the file. If the user does not enter any information for a field, `ADDHEAD` will use the default. The defaults for each field are printed with the prompt in parenthesis. After all information has been entered, `ADDHEAD` reads the binary file and writes the header and data to a new file called `filename.dat`.

A.6 REFERENCES

M68HC11EVBU Universal Evaluation Board User's Manual. 1986. Motorola Literature
    Distribution, P.O. Box 20912, Phoenix, AZ 85036.
ADA2100 User's Manual. 1990. Real Time Devices, Inc, 820 North University Dr., P.O. Box
    906, State College, PA 16804, 1990.
M68HC11 User's Manual. 1990. Motorola Literature Distribution, P.O. Box 20912, Phoenix,
    AZ 85036.

Appendix B

Data Acquisition and Control Routines

Danial J. Neebel

This appendix describes the routines that can be used to perform data acquisition and control provided in DACPAC.LIB. The routines in DACPAC were developed for two purposes. The first was for use in UW DigiScope. The second was to provide simple software interfaces to the signal conversion devices for use by other programs. DACPAC.LIB and the required headers are included on the floppy disk you received with this text. The DACPAC routines perform analog signal acquisition, digital signal acquisition, analog signal output, digital signal output, and timing.

The DACPAC routines provide an interface to three different devices. These three devices are an Analog and Digital I/O interface card from Real Time Devices that is installed inside the IBM PC or compatible, an external single board computer with RS232 communications (Motorola S68HC11EVBU Student Project Kit), and a virtual I/O device (data files).

There are some similarities between the three different devices. All device handling is done with four basic operations. These are OPEN, GET, PUT, and CLOSE. OPEN initializes the device. GET retrieves a piece of data from the device. PUT gives the device a piece of data to output. CLOSE terminates all operations being performed on or by the device, including closing a file, disabling interrupts, and removing communication links.

There are five sections in this appendix. In the first section we present the data structures used in DACPAC.LIB. The next section describes the top-level routines used to call the routines specific to each device. The following three sections discuss the routines provided in DACPAC. The descriptions of the routines for handling devices are divided up into subsections for each type of routine. Along with the four types above, we have added digital input and digital output as two special types of GET and PUT routines. Each subsection describes in detail what data are required by the routine and what data are set by the routine.

Before charging off into details, we give one simple warning. Always make sure to close all devices that have been opened before exiting the program. This means that the use of the C-library function exit() should be used with great care. A device left open after program exit could cause your PC to hang up.

B.1 DATA STRUCTURES

Here are the important data structures used by the routines described in this section. The most important is the `Header` data structure. This structure contains all the information about the data that has been or will be gathered from a device.

The title, creator, source, and type are all strictly character strings to be used as the programmer sees fit. The package routines do not use these fields. When opening any of the devices supported by DACPAC, the user must be careful to initialize some parts of the `Header` data structure passed to the OPEN routine and note also what parts of the data structure are initialized by the OPEN routine. Figure B.1 shows the data structures used by the data acquisition routines in DACPAC.

```
typedef short DATATYPE;      /* data type will be 16 bit integer */

typedef enum {ECG,EMG,EEG,CV,RESP,EKG,ABC,ERROR}  chantype_t;

typedef struct ChannelRecord {
    chantype_t    type;
    float         offset;
    float         gain;
} CHANTYPE;

typedef struct HeaderRecord {
    char      title[80];        /* title to be used for display */
    char      filename[40];     /* filename containing data */
    FILE      *fd;              /* file pointer returned by fADopen */
    char      creator[80];      /* name of person who gathered data */
    char      source[80];       /* name of A/D card or other device */
    char      type[80];         /* i.e. 12-lead ECG  */
    float     volthigh;         /* High limit of input voltage */
    float     voltlow;          /* Low limit of input voltage */
    int       stepsize;         /* step size used by data compress. */
    char      compression;      /* Data compression type */
    int       rate;             /* positive integer */
    int       resolution;       /* number between 8 and 16 */
    int       num_channels;     /* number of channels (in array) */
    int       num_samples;      /* number of samples */

    CHANTYPE channel[20];       /* pointer to array of Channel info*/
    DATATYPE     *data;         /* pointer to data buffer */
} DataHeader_t;
```

Figure B.1 Data header and channel type data structures (from `defns.h`).

An important part of the `Header` structure is the array of `CHANTYPE`. This array of structures contains all the information that is unique to each channel. Part of the

CHANTYPE structure is an enumerated typed variable called **type**. This field is not used by any of the **DACPAC** routines. The **type** field is provided for the user.

The defined type **DATATYPE** is more than just type short. The convention for all variables of type **DATATYPE** is that they be 2's complement 16-bit integers, a value of 0 corresponds to 0 V. To calculate the voltage given from a value of **DATATYPE**, multiply the value by the resolution (i.e., bits/V) of the data from device being read. To calculate the number of bits/V, divide the voltage range by 2 raised to the power of the resolution. The voltage range is found by subtracting **Header->voltlow** from **Header->volthigh**. The resolution is taken from **Header->resolution**.

B.2 TOP-LEVEL DEVICE ROUTINES

DACPAC contains routines that call specified device handlers. These routines are included in **DAC.C**. The header file that contains the prototypes for the routines is **DAC.H**. The prototypes are shown in Figure B.2. Examples of the calling conventions for each type of device are given in Figure B.3. For determining what values to send these routines, look at the section corresponding to the device you are using. The top-level routine **PUT()** calls **fADput_buffer()** with a size of one.

```
char OPEN(DataHeader_t *Header, char *dir, int device);
char CLOSE(DataHeader_t *Header, int device);
char GET(DataHeader_t *Header, DATATYPE *data, int device);
char PUT(DataHeader_t *Header, DATATYPE *data,int channel, int
device);
```

Figure B.2 Prototypes for the top level routines (from DAC.H).

B.3 INTERNAL I/O DEVICE (RTD ADA2100)

The RTD (Real Time Devices) ADA2100 can perform more I/O functions than the other two devices described in this appendix. Using the routines below, the ADA2100 is capable of reading eight single-ended analog inputs, driving two analog outputs ranging from –10 to +10 V, reading four digital inputs, and driving four digital outputs. The digital I/O uses standard TTL levels ranging from 0 to +5 V.

```
#define VIRTUAL    1
#define EXTERNAL   2
#define INTERNAL   3

/* calling conventions for Internal device */

OPEN(&Header, "", INTERNAL);
CLOSE(&Header, INTERNAL);
GET(&Header, &data, INTERNAL);

/* calling conventions for External device */

OPEN(&Header, "r", EXTERNAL);
CLOSE(&Header, EXTERNAL);
GET(&Header, &data, EXTERNAL);

/* calling conventions for Virtual device */

OPEN(&Header, "rt", VIRTUAL);
CLOSE(&Header, VIRTUAL);
GET(&Header, &data, VIRTUAL);
```

Figure B.3 Calling conventions for the three devices.

See Appendix A for a description of how to configure the ADA2100. Some items such as configuring for digital input and output are done via the software in DACPAC. Figure B.4 shows the routines provided to perform the above functions and the calling convention for each. The prototypes for all the routines shown in the figure are in IDAC.H. All routines return a value of 1 if the operation is successful and 0 if the operation is unsuccessful. Some routines will always be successful and will always return a value of 1.

```
char Iopen(DataHeader_t *Header);
char Iclose(DataHeader_t *Header);
char Iget(DATATYPE *data);
```

Figure B.4 Internal card calling conventions (from IDAC.H).

B.3.1 Opening the device: Iopen()

The Iopen() routine initializes the ADA2100 to perform digital input and output and analog input at the given sample rate. The timer is set to provide interrupts at the given sample rate. The interrupt vector is set to point to a small routine that sets a flag. The flag is checked by Iget().

Figure B.5 shows the how the Header data structure is used and what elements of the structure must be initialized and what elements are initialized by Iopen().

Any elements not listed are not initialized or used by `Iopen()` and so can be used at the discretion of the calling routine.

```
The following items must be initialized before the device is opened.

    Header->rate
    Header->num_channels

The following items are initialized by the Iopen(); routine.

    Header->volthigh = 5.0;
    Header->voltlow = -5.0;

    Header->resolution = 12;
    if (channels > 7) channels = 7; /* should fix this sometime */
    Header->num_channels = channels;
    Header->num_samples = 0;

    for (i=0;i<Header->num_channels;i++) {
        ChanOffset(Header,i) = 0.0;
        ChanGain(Header,i)  = 1.0;
    }
```

Figure B.5 `Iopen()` routine description.

B.3.2 Closing the device: Iclose()

The `Iclose()` routine stops the timer on the ADA2100 and resets the interrupt vector to the value set before `Iopen()` was called. `Iclose()` does not modify nor does it require any elements of the **Header** data structure. **Header** is passed into `Iclose()` only for commonality.

B.3.3 Taking an analog input reading: Iget()

The `Iget()` routine checks to see if the flag has been set by the timer interrupt routine. If the flag has been set, `Iget()` reads the channels requested by `Iopen()`, puts the data from those channels in an array of **DATATYPE** pointed to by data and returns a 1. If the flag has not been set , `Iget()` returns a 0. If a rate of 0 is selected, then `Iget()` will always read a value into data and return a 1.

B.4 EXTERNAL I/O DEVICE (MOTOROLA 68HC11EVBU)

The routines that interface to either of the Motorola devices do so by performing serial communications on an RS232 link. Both the EVBU and EVB use the same interface. The routines provided in DACPAC are given in Figure B.6. These are similar to the routines provided for the ADA2100. There are two major differences.

Although it is not absolutely necessary, we recommend to those using an EVBU for data acquisition and control that some protection be placed on the inputs and outputs. See section A.2.4 for some examples of simple protection circuits.

```
char Eopen(DataHeader_t *Header, char *dir);
char Eclose(DataHeader_t *Header);
char Eget(DATATYPE *data);
```

Figure B.6 Edac calling conventions (from EDAC.H).

B.4.1 Opening the device: Eopen()

The Eopen() routine initializes the serial port and sends a soft reset command followed by setup commands to the EVBU. These setup commands tell the software running on the EVBU at what rate to sample the analog inputs and which inputs are to be sampled. Since the EVBU can only communicate at a maximum of 9600 bits/s, it is necessary in some cases to buffer the data and send the block to the PC when the PC has time to read the data. So a last item included in these commands is whether real-time transfer or block transfer is to be used. To set the EVBU for real-time transfer, call Eopen() with dir set to r. To set the EVBU for block transfers, call Eopen() with dir set to b.

Figure B.7 shows how the Header data structure is used, what elements of the structure must be initialized, and what elements are initialized by Eopen().

```
The following items must be initialized before the device is opened.

    Header->rate
    Header->num_channels

The following items are initialized by the Eopen(); routine.

    Header->volthigh = 5.0;
    Header->voltlow = -5.0;

    Header->resolution = 8;
    if (channels > 8) channels = 8;
    Header->num_channels = channels;
    Header->num_samples = 0;

    for (i=0;i<Header->num_channels;i++) {
        ChanOffset(Header,i) = 0.0;
        ChanGain(Header,i)  = 1.0;
    }
```

Figure B.7 Eopen routine description.

Any elements not listed are not initialized or used by `Eopen()` and so can be used at the discretion of the calling routine.

If you do not wish to perform analog input, simply give `Eopen()` a sample rate of 0. If this is done, no analog inputs will be read. In this case `Eget()` should not be used. `Eopen()` assumes the caller does not want to perform analog input. Digital input and output may still be performed.

B.4.2 Closing the device: Eclose()

The `Eclose()` routine will stop all analog signal acquisition on the EVBU. The EVBU will be given a soft reset command. This means that the EVBU will stop reading analog inputs and performing digital I/O until `Eopen()` is called again and the EVBU receives setup information again.

B.4.3 Taking an analog input reading: Eget()

The `Eget()` routine checks to see if data has arrived from the EVBU. If data has arrived without error, then the `Eget()` puts the data from the channels in DATATYPE variable pointed to by data and returns a 1. If no data has arrived from the EVBU, `Eget()` returns a 0.

B.5 VIRTUAL I/O DEVICE (DATA FILES)

Figure B.8 shows the calling conventions for each of the routines available to the user. These routines were developed to serve two purposes. The first was to allow anyone to do real-time data acquisition even if they do not have any data acquisition hardware. The second purpose was to allow the storage, labeling, and retrieval of signals gathered using the above data acquisition routines. The routines described here do not require any special hardware. If you have an IBM PC/XT/AT or compatible with a floppy drive, 640 kbytes of RAM, and either a Hercules monochrome or VGA color monitor, you should be able to run these routines.

```
char fADopen(DataHeader_t *Header, char *dir);
char fADclose(DataHeader_t *Header);
int fADget_buffer(DataHeader_t *Header,int size,DATATYPE *buffer);
char fADget(DataHeader_t *Header, DATATYPE *signal);
int fADput_buffer(DataHeader_t *Header,int size,DATATYPE *buffer);
void fADputfile(DataHeader_t *Header);
void fADgetfile(DataHeader_t *Header);
```

Figure B.8 fAD calling conventions (from fAD.H).

B.5.1 File structure

The file structure uses an ASCII text header to describe the data in the file. The header is terminated by a single blank line. The data immediately follow this blank line. The data are stored in the file in binary. Figure B.9 gives an example `Header`. Note that for `fADopen()` to operate properly, the field names must be exactly as shown in Figure B.9. Even though the headers are stored in text format, the files cannot be edited using a normal text editor since the data are stored in the files in binary. Section A.5 shows how to use the program called `ADDHEAD` to add a header to an existing file.

```
Title:   ECG data from file ecg105
Creator:   unknown
Source:   unknown
Type:  ECG single channel
Volthigh:  12
Voltlow:   -12
Step:  0
Compress: N
Resolution:   12
Rate:  200
Channels: 1
Samples: 12000
Chan:  0    Gain   1.0000   Ofst   0.0000     Type ECG
```

Figure B.9 Example file header.

B.5.2 Opening the device: fADopen()

The `fADopen()` routine opens a file and initializes the timer to provide software delays to simulate real-time data acquisition. If you do not wish to perform timed analog input, simply call `fADopen()` with `dir` set to `r` for read only. If timed input is desired, then call `fADopen()` with `dir` set to `rt` for read with timed input. If file output is desired, then call `fADopen()` with `dir` set to `w` for write only. When opening a file for output, *all* elements of the `Header` structure must be set. When opening a file for read only `Header->filename` need be set. As shown in Figure B.10, all other elements will be read in from the file. Function `fADopen()` will only open a file for reading (options `r` and `rt`) or writing (option `w`). This figure shows how the `Header` data structure is used and what elements of the structure must be initialized and what elements are initialized by `fADopen()`. Any elements not listed are not initialized or used by `fADopen()` and so can be used at the discretion of the calling routine.

```
The following items must be initialized before the device is opened.

  Header->filename;

The following items are initialized by the Iopen(); routine.

  Header->title;
  Header->creator;
  Header->source;
  Header->type;
  Header->fd;
  Header->stepsize;
  Header->compression;
  Header->rate;
  Header->resolution;
  Header->num_samples;
  Header->volthigh;
  Header->voltlow;
  Header->compression;        /* no compression */
  Header->resolution;
  Header->channels = channels;
  Header->num_samples = 0;

  for (i=0;i<Header->num_channels;i++) {
     ChanOffset(Header,i);
     ChanGain(Header,i);
     ChanType(Header,i);
  }
```

Figure B.10 `fADopen` routine description.

B.5.3 Closing the device: fADclose()

The `fADclose()` routine performs two very important operations. First, the obvious, it closes the file pointed to by `Header->fd`. Second, if `fADopen()` was called with `dir` pointing to `t`, `fADclose()` will remove the timer routine from the time-of-day interrupt line.

B.5.4 Taking an analog input reading: fADget(), fADget_buffer(), fADgetfile()

The `fADget()` routine checks the flag set by the interrupt service routine. If the flag has been set, then `fADget()` reads one sample from each of the channels in the file pointed to by `Header->fd` and returns a 1. The samples are then placed in an array pointed to by data. It is the responsibility of the calling routine to allocate enough space for the data. If the flag has not been set by the interrupt service routine, then `fADget()` returns a 0.

If no timing is required and all of the data needs to read in for processing, then `fADgetfile()` can be used. `fADgetfile()` will open the file specified by `Header-`

`>filename`, read the file header into the `Header` data structure, and read all the data into an array pointed to by `Header->data`. Note also that `fADgetfile()` will allocate memory for the data pointed to by `Header->data`.

If the file is too large to read all at once or only some of the data is needed from the file, then `fADget_buffer()` is very useful. `fADget_buffer()` will read as many samples as are requested by size. If successful, `fADget_buffer()` returns the number of samples read from the file. A sample is one reading on each channel.

B.5.5 Writing data to a file: fADput_buffer() and fADputfile()

For writing data out to a file, there are two routines—`fADput_buffer()`, and `fADputfile()`. The routines used for writing to a file do not use timing. There is no `fADput()` to correspond with `fADget()` since calling `fADput_buffer()` with a buffer size of 1 will write one sample of each channel into the file pointed to by `Header->fd` just as one would expect a routine called `fADput()` to do. Remember that a sample is one reading for each channel. The top-level routine `PUT()` calls `fADput_buffer()` with a size of one.

`fADputfile()` makes storing the data to a file simple. `fADputfile()` will open the file name `Header->filename` for writing, write all header information to the file, and write all data to the file and then close the file. `fADputfile()` does not free the memory pointed to by `Header->data`.

B.6 REFERENCES

ADA2100 User's Manual. 1990. Real Time Devices, Inc, 820 North University Dr., P.O. Box 906, State College, PA 16804. Ph: (814) 234-8087; Fax: (814) 234-5218.

Eggbrecht, L. C. 1983. *Interfacing to the IBM Personal Computer.* Indianapolis, IN: Howard W. Sams.

M68HC11 User's Manual. 1990. Motorola Literature Distribution, P.O. Box 20912, Phoenix, Arizona 85036.

M68HC11EVB Evaluation Board User's Manual. 1986. Motorola Literature Distribution, P.O. Box 20912, Phoenix, AZ 85036.

Turbo C Reference Guide. 1988. Borland International, Inc., 1800 Green Hills Road, P.O. Box 660001, Scotts Valley, CA 95055-0001.

Turbo C User's Guide. 1988. Borland International, Inc., 1800 Green Hills Road, P.O. Box 660001, Scotts Valley, CA 95055-0001.

Appendix C

Data Acquisition and Control—Some Hints

Danial J. Neebel

In the previous two appendices we described how to set up the hardware to use with UW DigiScope and how some of the routines in DACPAC can be used. Here we provide some helpful hints on how to develop some of these routines and set up the hardware to perform DAC. We are not describing everything that must be done. We give the reader some hints and advice as to what to do and what not to do. We will also give some good references for performing some of the operations needed to do data acquisition and control with the IBM PC.

This appendix should give the reader some idea of how to go about setting up a simple data acquisition and control system. It includes information on the basic elements required to read analog and digital signals into a computer using the three different types of devices presented in the previous appendix. The first type is an internal device. An internal device is connected to and communicates with the PC via the internal expansion bus. An I/O device can also be external. An external device communicates with the PC via either a serial or parallel communication port. The most common communication ports are RS232 (serial) and IEEE 488 (parallel). In this text, we discuss only RS232 communications since almost all IBM PC architecture machines have an RS232 serial port available. The last type of device is a virtual I/O device. DigiScope uses data files and a timer interrupt to simulate analog data acquisition. These same file utilities are used to store and retrieve data gathered using internal and external analog input devices.

There are four basic operations involved in using an input/output device: Open Device, Input Data, Output Data, and Close Device. Open Device will initialize the device for the type of I/O requested and set up any interrupts that may be needed to perform exact timing. Input Data will determine if the data requested is available and retrieve the data from the device. Output Data will give the device a piece of data to output. Close Device will terminate all operations being performed by the device and disable any interrupts set up by Open Device. In the discussions that follow, we describe exactly what must be done to perform each of these operations for each of the three devices presented. The last section gives helpful hints for writing your own interface to be used with DigiScope.

All programming for the PC is done in Turbo C. We refer the reader to the Turbo C manuals for information on such things as serial communication routines and file input/output.

C.1 INTERNAL I/O DEVICE (RTD ADA2100)

This type of device requires installation inside the chassis of the IBM PC. Figure C.1 shows the connection of an ECG amplifier to an I/O card. Note that no extra hardware is required.

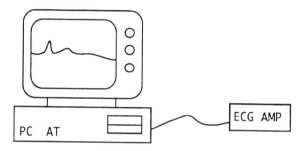

Figure C.1 Internal card data acquisition and control system.

C.1.1 Interfacing to an I/O card with Turbo C

Interfacing to a card installed in the PC is done using the library routines provided in Turbo C that read and write from and to the I/O space of the processor. For 8-bit I/O operations, `inportb()` and `outportb()` are used for input and output respectively. Functions `inport()` and `outport()` are used for 16-bit input and output. The ADA2100 is an 8-bit I/O card so we have used `inportb()` and `outportb()`. We refer the reader to the Turbo C Reference Guide for more information on these routines. Figure C.2 shows examples of using `inportb()` and `outportb()` to set up the 8259 interrupt controller. The 8259 is part of the PC system, but the operations of reading and writing using `inportb()` and `outportb()` are the same as reading and writing to a card.

C.1.2 Handling interrupts on the IBM PC with Turbo C

We discuss some of the basic operations required to properly set up interrupts and—possibly more important—how to make sure that interrupts are disabled when we do not want them to occur. Programming with interrupts is difficult because an interrupt can occur at any point in the execution of a program. We have limited control over when an interrupt can occur. An interrupt is used when an external event needs to stop whatever process is running and cause another process to execute. There are many sources of interrupts. An interrupt can come from the keyboard, a disk controller, a serial port, the time of day interrupt, or many other sources. In this case we would like something to happen at very regular time intervals.

Interfacing to the interrupt controller

The IBM PC has one Intel 8259 interrupt controller. The interrupt controller is located at 0x20 and 0x21 in the I/O space of the processor. Note that we use the C-language convention for specifying hexadecimal (base 16) constants. The interrupt controller is the device that tells the processor that an external process has requested an interrupt. If interrupts are enabled, the processor will acknowledge the interrupt. The interrupt controller will then tell the processor where to look for the interrupt vector. The vector is the address of the interrupt service routine.

The 8259 interrupt controller has many capabilities, but we recommend that you only change the interrupt mask register, IMR. For a more detailed description of the 8259, see the Intel *Microprocessor and Peripheral Handbook, Volume I*, or Eggbrecht (1983). The IMR is located at 0x21 in I/O space. The IBM PC/AT architecture uses two 8259 interrupt controllers. The master interrupt controller is located at 0x21 and the slave is located at 0x70. The two interrupt controllers are cascaded to provide 15 different interrupt levels. IRQ2 of the master interrupt controller is connected to the slave.

When masking or unmasking an interrupt, it is very important to only change the mask of the interrupt of interest. It is equally important to mask the interrupt after use. Figure C.2 shows one method of unmasking an interrupt at the beginning of a program and returning the mask to the original setting at the end of the program.

Interfacing to the operating system—interrupt handling

Along with unmasking the interrupt in the interrupt controller, we must also initialize the interrupt vector to point to the proper interrupt handler. Turbo C provides two routines that make this very easy. Before setting the interrupt vector to the new interrupt handler, the current interrupt vector must be saved so that before the program exists to the operating system, the interrupt vector can be returned to the original value.

An important item to remember when using interrupts on the PC is to always return the system interrupt handler and interrupt vectors to the state they were in when the program started. To do this we save the current IMR and interrupt vector into two global variables and reset the IMR and interrupt vector to these values upon termination of the program. Exiting without resetting the IMR and interrupt vector could cause serious problems. If this event occurs for whatever reason, the user should reboot the system.

```
main(char *argv[], int argc)

{
  disable();                        /* disables all interrupts */

  OldVector = getvect(10);          /* set interrupt vector */
  setvect(10, Itimer);

  OldMask = inportb(0x21);          /* unmask IRQ 2 on 8259  */
  mask = OldMask & 0xFB;
  outportb(0x21,mask);
  enable();                         /* enables all interrupts */

/*  code that can be interrupted by INTERRUPT 10  */

  disable();

  outportb(0x21,OldMask);           /* return mask to original value */
  setvect(10, OldVector);           /* return interrupt vector to    */
                                    /* original value                */
  enable();
}
```

Figure C.2 Setting the interrupt vector and interrupt mask register.

C.2 EXTERNAL I/O DEVICE (MOTOROLA 68HC11EVBU)

For an external input/output device, we have chosen the Motorola S68HC11EVBU student project kit. This device was chosen because of its availability and low cost. The drawback of using this device is the difficulty in setting it up. To turn the EVBU into an I/O device, a 68HC11 assembly language program is needed to communicate with the PC and perform the necessary input and output. Fortunately we have done this part of the task for you. An example system is shown in Figure C.3. Also, the EVBU is not a difficult device to program. The student evaluation kit includes enough tools and documentation so that anyone who has experience writing assembly language code should be able to program the EVBU.

An advantage of using this device is that all critical timing can be done by the 68HC11. This means we can avoid using interrupts on the PC altogether. However, the 68HC11 has a time-based interrupt. Interrupts are a little easier to handle on a microcontroller than on a PC; an operating system and disk drives and other items make a PC more complicated.

In this section we give a short introduction to using serial communications with Turbo C. Finally, we give an example of two routines, one written in Turbo C for the PC and one in assembly language for the EVBU; they each perform a simple communication sequence.

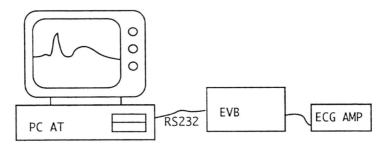

Figure C.3 Motorola EVBU data acquisition and control system.

C.2.1 Common operations

Here we briefly describe the operations needed to perform the basic operations of
OPEN, **CLOSE**, **GET**, and **PUT**.

```
#define COM1      0
#define COM2      1
#define COM3      2
#define COM4      3
define   SETTINGS  (0xE0|0x03|0x00|0x00)
                   /* 9600 Bd 8 bits 1 stop no parity */

                   /* CPORT contains COM port is being used */
int   CPORT=COM1;

                   /* initialize serial port */
bioscom(0, SETTINGS, CPORT);

                   /* read the status of serial port */
status = bioscom(3,0,CPORT);

                   /* read the input buffer of serial port */
port = bioscom(2,0,CPORT);

                   /* Send an ASCII Q out on serial port */
bioscom(1,'Q',CPORT);
```

Figure C.4 Use of the **bioscom()** Turbo C library routine.

Opening the device

An open device routine must initialize the RS232 serial port, establish a communication link with the EVBU, and initialize the EVBU for any timing and data taking

that need to be performed. Initializing the serial port is easily done using the
`bioscom()` routine provided by Turbo C (see Figure C.4). In section C.3.3 we give
an example of how this is done. Figure C.5 shows the communication sequence
that takes place when the external device is opened.

```
- PC sends reset command "R" to EVBU for soft reset.
- EVBU gets "R" and jumps to reset vector. As part of reset, EVBU
  sends "R" to PC to echo Reset command.
- PC sends channel mask "Cx" to EVBU. x is an 8-bit mask with 1's
  in positions corresponding to analog input channels to be read.
- EVBU echoes channel mask "Cx" and saves it.
- PC sends delay setting, "Dxxxx" to EVBU. xxxx is in Hex and is
  timer ticks of 68HC11 timer.
- EVBU echoes "Dxxxx" command and saves delay setting.
- PC sends go command "G" to EVBU.
- EVBU echoes "G" command and starts timer interrupts.
```

Figure C.5 Communication sequence for initializing EVBU for real-time data transfer.

Closing the device

A close device routine need only send a command to tell the EVBU to terminate all
data acquisition and stop sending data to the PC host.

Important: Starting EDAC

Each time the EVBU is reset it is necessary to use
`pcBug11` to restart the EDAC program. The command to do
this is:

PCBUG11 -E port=N macro=go.

Where **N** is the number of the COM port that the EVBU
is connected to. **N** must be either 1, 2, 3, or 4. Also the
following files must be in the current directory:

PCBUG11.EXE, TALKE.XOO, TALKE.BOO, and GO.MCR.

All these files except for **GO.MCR** are included on the
`pcBug11` disk you received with the EVBU. **GO.MCR** is
included on your DigiScope disk.

Input data

An input data routine must give the EVBU a request to read data and wait for a response. Remember that the host should always limit the amount of time spent waiting for a response so as not to hang the computer waiting for an event that will never occur.

Output data

An output data routine must send the EVBU a command telling the EVBU to output a specified piece of data.

C.2.2 Serial communications using Turbo C

Serial communication is accomplished via the **bioscom()** routine. Function **bioscom()** is a multipurpose routine that can be used to initialize the serial port, check status, read the serial port, send a byte out on the serial port. The Turbo C reference manual gives some very helpful examples of how to perform each of these operations. Figure C.6 shows some examples from the code developed for this text.

```
  send_command("R");                    /* reset the EVBU */

/* Tell EVBU which channels the host wants to read. */
/* The EVBU will send this many bytes to the TERMINAL */
/* at the sample rate */

  for (i=0, mask=0;i<Header->num_channels;i++) {
     mask |= 1 << i;
  }
  sprintf(buf,"C%X\r",mask);
  send_command(buf);

                              /* calc delay betw. samples */
  if (Header->rate != 0) {
    delay = 2000000 / Header->rate;
  } else {
    delay = 0;                 /* if rate is 0  send 0 delay */
  }
  sprintf(buf,"D%X\r",delay);      /* Tell EVBU delay */
  send_command(buf);

  send_command("G");          /* Tell EVBU to start sampling */
```

Figure C.6 Turbo C code for the PC to execute the sequence in Figure C.5.

C.2.3 A sample communication sequence and the code to execute it

Here we present the communication sequence used to open the EVBU device used by Eopen(). We have removed some of the error checking to make the code easier to understand. The sequence being executed is exactly the same as that performed by Eopen() on the PC and EDAC.ASM on the EVBU. The sequence is outlined in Figure C.5. The PC must provide the EVBU with sample rate and number of channels, and indicate if block transfer mode or real time mode is to be used. The code used to execute the sequence on the PC and EVBU is given in Figures C.6 and C.7 respectively. The send_command() routine shown in Figure C.8 sends a NULL-terminated string to the EVBU.

Note that a one-byte mask is sent to tell the EVBU which channels to read. Each bit position corresponds to an analog input channel. The bits and channels are numbered 0–7. If bit 2 is the only bit set in the mask, then channel 2 will be the only channel read.

```
        LDAA    #'R'              send signon character to host
        JSR     DATOUT
SETUP   JSR     DATIN
        BRCLR   FLAGS
        RCVDAT  SETUP         wait for a character
        JSR     DATOUT        echo to host for host's error
                                checking
        CMPA    #'D'          Check for delay
        BNE     SETUP1
        JSR     GETDELAY      Read in the hex value and save
        BRA     SETUP
SETUP1  CMPA    #'C'          Check for number of channels
        BNE     SETUP2
        JSR     GETCHAN       Read in the hex value and save
        BRA     SETUP

SETUP3  CMPA    #'G'          Check for Start
        BNE     SETUP

* set up A/D converter and start interrupts
```

Figure C.7 Assembly language code for 68HC11 to execute the sequence in Figure C.5.

C.3 VIRTUAL I/O DEVICE (DATA FILES)

File I/O is a normal operation to most programmers. Here we show one method of using files to simulate a physical I/O device. The first task is to make file I/O operations look like operations involving a physical I/O device. Next we need to provide some type of timing operation so that input and output take place at a specific rate. Providing timing is the difficult part of simulating a physical device with data files. To perform the timing operation of virtual analog input, your system must have the time-of-day interrupt compatible with an IBM PC or PC/AT.

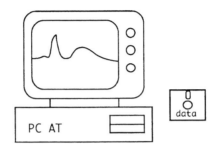

Figure C.8 A microcomputer and hardware for virtual analog input.

C.3.1 Stealing the time of day interrupt

Figure C.9 shows how to steal the time-of-day interrupt. The old interrupt vector is saved so that it can be restored and also so that the new timer interrupt routine can call the interrupt service routine approximately 18.2 times/s. This keeps the time-of-day clock running at a rate close to the correct rate. If this was not done the time-of-day would be set to an unknown value after the program was finished using the interrupt. For more information about what types of things can be done by stealing the time of day interrupt, see Bovens and Brysbaert (1990).

C.3.2 Initializing the 8253 for a specified sample rate

Figure C.10 shows a routine to set up the timer to a specified rate. The operations required are to calculate the number of clock ticks required between samples and then set up the timer. Timer setup requires setting **TIMER0** in **mode 2** and writing the least-significant byte then the most-significant byte to **TIMER0**. The timer control is located at 0x43 in I/O space and **TIMER0** is located at 0x40 in I/O space. The 8253 has three timers. **TIMER1** is used for the speaker output and **TIMER2** is used for dynamic memory refresh. It is very important that both of these timers remain undisturbed. We refer the reader to the Intel *Microprocessor and Peripheral Handbook, Volume II* for more information on the Intel 8253.

```
#define TIMER0            0x40        /* I/O mem locations for 8253 */
#define TIMER_CTRL        0x43

void interrupt(*OldTimer)(void); /* global to save old ISR */

  disable();                              /* disable interrupts */
  OldTimer = getvect(0x08);               /* save the old ISR*/
  setvect(0x08, SuperTimer);              /* set new ISR to our routine */

  SetUpTimer(rate);                       /* initialize timer to rate */

  enable();                               /* enable interrupts */

/***** code that can be interrupted goes here ******/

  disable();                              /* disable interrupts */
  setvect(0x08, OldTimer);                /* restore ISR */

  outportb(TIMER_CTRL, 0x36);    /* timer 0,mode 3, LSB and MSB */
  outportb(TIMER0, 0x00);
  outportb(TIMER0, 0x00);

  enable();                               /* enable interrupts */
```

Figure C.9 Code to steal the time-of-day interrupt (INT 8).

```
#define TIMER0            0x40        /* I/O mem locations for 8253 */
#define TIMER_CTRL        0x43

#define TIMER_CLOCK       (long)1192755   /* Hz crystal for 8253 */
#define BIOS_TIC          (double)18.2    /* in ticks per second */

int old_timer_call;

void SetupTimer(frequency)
int  frequency;
{
  long                   divisor;
  int   data;

  old_timer_call = (int) (((double) frequency) / BIOS_TIC);

  divisor = TIMER_CLOCK / ((long) frequency);

  outportb(TIMER_CTRL, 0x34);/* timer 0, mode 3, LSB and MSB */
  data = (int) (divisor & 0xFF);
  outportb(TIMER0, data);
  data = (int) ((divisor >> 8) & 0xFF);
  outportb(TIMER0, data);
}
```

Figure C.10 Code to set the 8253 TIMER0 to for a specified sample rate.

C.3.3 Writing an interrupt service routine in Turbo C

It is important that an ISR return the system to the original state the system was in before the interrupt service routine was executed. This means that an ISR must save the current state of the processor when the ISR was started. Fortunately, Turbo C makes sure this is done if the routine is declared with "void interrupt."

```
void interrupt SuperTimer()
{
  static int  super_counter = 0;

  TIME_OUT = TRUE;

  if (++super_counter == old_timer_call)
  {
     OldTimer();             /* execute old timer approx. 18.2 */
     super_counter = 0;      /* times per second */
  } else {
     outportb(0x20, 0x20); /* interrupt acknowledge signal */
  }
}
```

Figure C.11 Interrupt service routine to set flag and call real-time clock approximately 18.2 times/s.

Figure C.11 shows an ISR that could be used for virtual A/D in **DACPAC.LIB**. Note that the **SuperTimer()** routine only sets a flag and updates the real-time clock. By limiting the amount of work done by an ISR, we can eliminate some of the problems caused by using interrupts.

C.4 WRITING YOUR OWN INTERFACE SOFTWARE

One of the most common needs of users of the data acquisition software available with DigiScope will be adding a new interface device. In this section, we give some directions for adding an interface to a different internal card to DigiScope.

C.4.1 Writing the interface routines

Section C.2 gives some hints for one method of interfacing Turbo C code to an internal card. Most I/O cards will be shipped with some interface routines and/or a manual and examples for writing these routines. The trick to making the internal card work with DigiScope will be to match the card interfaces to the interfaces to DigiScope. Appendix B shows all of the function prototypes used in DigiScope for interfacing to the various types of I/O devices. For example, if you wish to write an

interface to an internal card, you will need to write the routines included in
IDAC.OBJ. A list of those functions is given in Figure B.4.

Examples of the code included in the module IDAC.OBJ. are given in Figures
C.12(a), C.12(b), and C.13. Figure C.12 shows the code used to open the device
and set up necessary timing and I/O functions. You should take note of the items in
the internal data structure **Header** that are initialized.

```
/* Global Variables  */

unsigned int   BaseAddress = 200; /* base address for RTD card */
unsigned int   IRQline = 3;
static  char   AD_TIME_OUT;
static  char   DA_TIME_OUT;
static  int    OldMask;
static  int    NUM_CHANNELS;       /* used to remember how many */
                                   /* channels to read */
static  char   OPEN=NO;
static  char   TIMED=NO;

/* RTD board addresses */

#define      PORTA         BaseAddress + 0
#define      PORTB         BaseAddress + 1
#define      PORTC         BaseAddress + 2
#define      PORT_CNTRL    BaseAddress + 3

#define      SOC12         BaseAddress + 4
#define      SOC8          BaseAddress + 5
#define      AD_MSB        BaseAddress + 4
#define      AD_LSB        BaseAddress + 5

#define      DA1_LSB       BaseAddress + 8
#define      DA1_MSB       BaseAddress + 9
#define      DA2_LSB       BaseAddress + 0xA
#define      DA2_MSB       BaseAddress + 0xB
#define      UPDATE        BaseAddress + 0xC
#define      CLEAR_DA      BaseAddress + 0x10

#define      TIMER0        BaseAddress + 0x14
#define      TIMER1        BaseAddress + 0x15
#define      TIMER2        BaseAddress + 0x16
#define      TIMER_CNTRL   BaseAddress + 0x17

char Iopen(DataHeader_t *Header)

{
  int register  i,j,in, out;
  int   delay, num_channels;
  char buf[20];
  int mask;
```

Figure C.12(a) Beginning of routine to open internal card for analog I/O.

```
  AD_TIME_OUT = FALSE;
  DA_TIME_OUT = FALSE;

  Header->volthigh = 5.0;   /* initialize header data structure */
  Header->voltlow = -5.0;

  Header->resolution = 12;
  if (Header->num_channels > 8)
     Header->num_channels = 8;
  NUM_CHANNELS = Header->num_channels;
  Header->num_samples = 0;
  Header->data = NULL;
/* initialize internal card  */
/* PORT C low is input */
/* PORT C high is output */

  outportb(PORT_CNTRL,0x91);

/* set gain and offset for each channel during Iget routine */
/* type should be set by user */
/* for now the gain is always set to one */

  for (i=0;i<Header->num_channels;i++) {
     ChanOffset(Header,i) = 0.0;
     ChanGain(Header,i) = 1.0;
  }

  if (!CheckTimer()) return(NO);    /* This routine is used to */
                              /* check if the board is installed */

                                     /* setup timing function */
  if (Header->rate != 0) {
     disable();                 /* only perform timing if rate is */
                                            /* nonzero */
     ISetupTimer(Header->rate);         /* setup 8253 timer */

     OldVector = getvect(IRQnumber);  /* set interrupt vector */
     setvect(IRQnumber, Itimer);
                                     /* unmask interrupt mask */
     OldMask = inportb(0x21);
     mask = OldMask & ~(1 << IRQline) & 0xFF;
     outportb(0x21,mask);
     enable();
     TIMED=YES;
  } else {                    /* rate is zero => don't use timing */
     TIMED=NO;
  }
  OPEN=YES;
  return(YES);
} /* Iopen() */
```

Figure C.12(b) End of routine to open internal card for analog I/O.

```
char Iget(DATATYPE *data)

{
    int msb, lsb, i;

    if (!OPEN) {
        DEVICE_NOT_OPEN();
        return(NO);
    }
    if (AD_TIME_OUT || !TIMED) {

        for (i=0;i<NUM_CHANNELS;i++) {
                /* read all channels requested  */
                /* for now the gain is tied to 1 */
        outportb(PORTB,0x00 | i );        /* select channel */
        outportb(SOC12,0);                /* start a conversion */
        while (!(inportb(PORTA) & 0x80));
                /* wait for conversion to end */
                /* after EOC then read MSB and LSB */
        msb = inportb(AD_MSB)*16;  /* read in data from RTD card */
        lsb = inportb(AD_LSB)/16;/* the card should be in +/- mode */
        data[i] = msb + lsb - 2048;
        }

        AD_TIME_OUT = FALSE;      /* Clear AD_TIME_OUT flag */
        return(YES);
    } else {
        return(NO);
    }
} /* Iget */
```

Figure C.13 Routine to get analog input data from internal card.

C.4.2 Including the new interface in UW DigiScope

The file **DACPAC.LIB** contains all of the UW DigiScope data acquisition routines. To replace the current interface for internal I/O card, you need only replace the **IDAC.OBJ** module in **DACPAC.LIB**. To replace the external device and virtual device, replace the **EDAC.OBJ** and **FAD.OBJ** modules respectively. You should include all of the functions that are listed as included with those modules in Appendix B. Failure to include all the functions will at the least cause a compiler error and at worst a run-time error. If you do not wish to use a function, you may simply insert a dummy function in its place. For additional details, see the **README** file on the UW DigiScope disk.

C.5 REFERENCES

ADA2100 User's Manual. 1990. Real Time Devices, Inc., 820 North University Dr., P.O. Box 906, State College, PA 16804.

Bovens, N. and Brysbaert, M. 1990. IBM PC/XT/AT and PS/2 Turbo Pascal timing with extended resolution. *Behavior Research Methods, Instruments, & Computers,* **22**(3): 332–34.

Eggbrecht, L. C. 1983. *Interfacing to the IBM Personal Computer.* Indianapolis, IN: Howard W. Sams.

M68HC11 User's Manual. 1990. Motorola Literature Distribution, P.O. Box 20912, Phoenix, Arizona 85036.

M68HC11EVB Evaluation Board User's Manual. 1986. Motorola Literature Distribution, P.O. Box 20912, Phoenix, AZ 85036.

Microprocessor and Peripheral Handbook, Volumes I and II. 1988. Intel Literature Sales, P.O. Box 8130, Santa Clara CA, 95052-8130.

Turbo C Reference Guide. 1988. Borland International, Inc., 1800 Green Hills Road, P.O. Box 660001, Scotts Valley, CA 95055-0001.

Turbo C User's Guide. 1988. Borland International, Inc., 1800 Green Hills Road, P.O. Box 660001, Scotts Valley, CA 95055-0001.

Appendix D

UW DigiScope User's Manual

Willis J. Tompkins and Annie Foong

UW DigiScope is a program that gives the user a range of basic functions typical of a digital oscilloscope. Included are such features as data acquisition and storage, sensitivity adjustment controls, and measurement of waveforms. More important, this program is also a digital signal processing package with a comprehensive set of built-in functions that include the FFT and filter design tools. For filter design, pole-zero plots assist the user in the design process. A set of special advanced functions is also included for QRS detection, signal compression, and waveform generation. Here we concentrate on acquainting you with the general functions of UW DigiScope and its basic commands. Before you can use UW DigiScope, you need to install it on your hard disk drive using the INSTALL program. See the directions on the DigiScope floppy disk. Also be sure to read the README.DOC file on the disk for additional information about DigiScope that is not included in this book.

D.1 GETTING AROUND IN UW DIGISCOPE

To run the program, go to the DIGSCOPE directory and type SCOPE. Throughout this appendix, the words SCOPE, DigiScope, and UW DigiScope are used interchangeably.

D.1.1 Main Display Screen

Figure D.1 shows the main display screen of SCOPE. There is a menu on the left of the screen and a command line window at the bottom. There are two display channels. The one with the dashed box around it is called the active channel, which can be selected with the (A)ctive Ch menu command. Operations generally manipulate the data in the active channel. The screen shows an ECG read from a disk file displayed in the top channel and the results of processing the ECG with a derivative algorithm in the bottom (i.e., active) channel.

Maneuvering in SCOPE is accomplished through the use of menus. The UP and DOWN ARROWS move the selection box up and down a menu list. Hitting the RETURN key selects the menu function chosen by the selection box. Alternately, a menu item can be selected by striking the key indicated in parenthesis for each command (e.g., key F immediately executes the (F)ilters command).

An e(X)it from the current menu to its parent (i.e., the previous menu) can be achieved either by placing the box on the e(X)it item and striking the RETURN key, by hitting the X key, or by simply hitting the ESC key. In fact the ESC key is used throughout SCOPE to exit from the current action, and the RETURN key is used to execute a selected function.

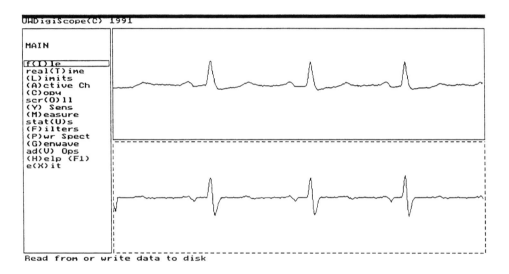

Figure D.1 UW DigiScope's main display screen. The dashed box indicates the active channel. In this case, the ECG in the top channel was read from a disk file, then a derivative algorithm was applied to processed the ECG and produce the waveform in the bottom (active) channel.

D.1.2 Communicating with SCOPE

Sometimes it is necessary for the user to enter data via the keyboard. Such data entry is done in the command line window at the bottom of the screen. Entry of data either ends with a RETURN, upon which the data are accepted, or an ESC which allows the user to quit data entry and make an escape back to the previous screen. Correction can be done with the BACKSPACE key prior to hitting RETURN. SCOPE also provides information to the user via short text displays in this window.

D.2 OVERVIEW OF FUNCTIONS

Figure D.2 shows how main menu functions branch to other menus. Command f(I)le permits reading or writing disk data files. Function real (T)ime lets you select the source of sampled data to be either from a disk file or, if the computer has the proper hardware installed, from an external Motorola 68HC11 microcontroller card or an internal Real Time Devices signal conversion card.

Function (L)imits lets you choose whether SCOPE functions operate on the whole file or just the portion of the file that is displayed. The default limits at start-up are the 512 data points from the file seen on the display. The maximal file size is 5,120 sampled data points. When you write a file to disk using the f(I)le (W)rite command, you are asked if you want to write only the data on the display (512 points) or the whole file. With the scr(O)ll function, you can scroll through a file using the arrow keys and select which of the file's 512 data points appear on the display. (C)opy performs a copy of one display channel to the other.

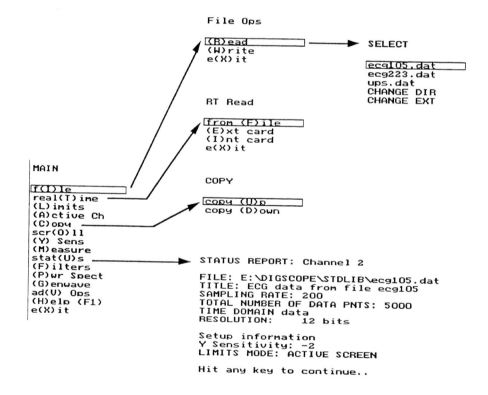

Figure D.2 Branches to submenus from the main menu.

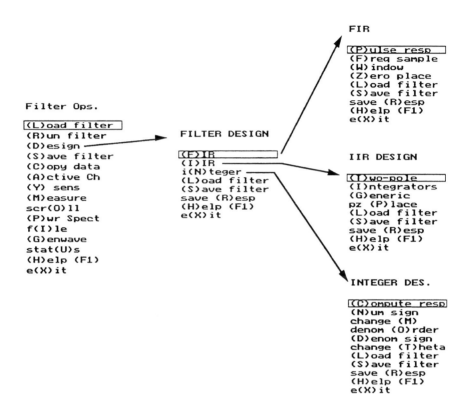

Figure D.3 Filter design menus.

To adjust the amplitude of the active channel, select (Y) Sens and increase or decrease the sensitivity of the channel by a factor of two each time you strike the up or down arrow on the keyboard. This function operates like the sensitivity control on an oscilloscope. (M) easure superimposes two cursors on the waveform in the active channel that you can move with the arrow keys. At the bottom of the display, a window shows the time and amplitude values of the cursors.

Command stat (U) s provides a summary screen of information about the characteristics of the current data display. This information is recorded in the header of a data file when it is written to disk.

Function (P) wr Spect computes and displays the power spectrum of a signal. A (H) elp function briefly explains each of the commands. Selecting to e (X) it from this main menu returns you to DOS. Choosing e(X)it on any submenu returns you to the previously displayed menu.

Figure D.3 shows menus for performing filter design that branch from the (F)ilters command in the main menu. Any filter designed with these tools can be saved in a disk file and used to process signals. Tools are provided for designing the three filter classes, FIR, IIR, and integer-coefficient filters. (F)IR and (I)IR each provide four design techniques, and I(N)teger fully supports development of this special class of filter. The most recently designed filter is saved in memory so that selecting (R)un filter executes the filter process on the waveform in the active window. (L)oad filter loads and runs a previously designed filter that was saved on disk.

Figure D.4 shows the filter design window for a two-pole IIR filter. In this case, we first selected bandpass filter from a submenu and specified the radius and angle for placement of the poles. The SCOPE program then displayed the pole-zero plot, response to a unit impulse, magnitude and phase responses, and the difference equation for implementing the filter (not shown). By choosing e(X)it or by hitting ESC, we can go back to the previous screen (see Figure D.3) and immediately execute (R)un filter to see the effect of this filter on a signal.

Note that the magnitude response is adjusted to 0 dB, and the gain of the filter is reported. High-gain filters or cascades of several filters (e.g., running the same filter more than once or a sequence of filters on the same signal data) may cause integer overflows of the signal data. These usually appear as discontinuities in the output waveforms and are due to the fact that the internal representation of signal data is 16-bit integers (i.e., values of approximately ±32,000). Thus, for example, if you have a 12-bit data file (i.e., values of approximately ±2,000) and you pass these data through a filter or cascade of filters with an overall gain of 40 dB (i.e., an amplitude scaling by a factor of 100), you produce numbers in the range of ±200,000. This will cause an arithmetic overflow of the 16-bit representation and will give an erroneous output waveform. To prevent this problem, run the special all-pass filter called atten40.fil that does 40-dB attenuation *before* you pass the signal through a high-gain filter. The disadvantage of this operation is that signal bit-resolution is sacrificed since the original signal data points will be divided by a factor of 100 by this operation bringing a range of ±2,000 to ±20 and discarding the least-significant bits.

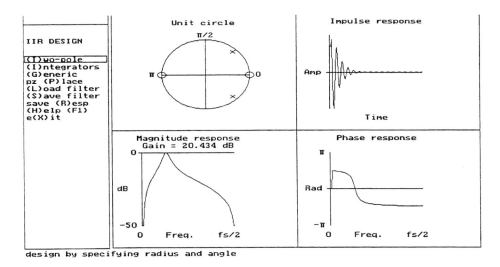

IIR DESIGN

(T)wo-pole
(I)ntegrators
(G)eneric
pz (P)lace
(L)oad filter
(S)ave filter
save (R)esp
(H)elp (F1)
e(X)it

design by specifying radius and angle

Figure D.4 UW DigiScope screen image.

Figure D.5 shows the set of special functions called from the main menu with the advanced options selection ad(V) Ops. These advanced features are frequency analysis, crosscorrelation, QRS detection, data compression, and sampling theory.

(F)req anal does power spectral analysis of a waveform segment (called a template) selected by the use of two movable cursors. This utility illustrates the effects of zero-padding and/or windowing of data. A template is selected with the cursors, and zero-padded outside of the cursors. Any dc-bias is removed from the zero-padded result. A window can be applied to the existing template. The window has a value of unity at the center of the template and tapers to zero at the template edges according to the chosen window. Function re(S)tore recopies the original buffer into the template.

(C)orrelation crosscorrelates a template selected from the top channel with a signal. A template selected from the upper channel is crosscorrelated with the signal on the upper channel, leaving the result in the lower channel. If you do not read in a new file after selecting the template, then the template is crosscorrelated with the file from which it came (an autocorrelation of sorts). After a template has been selected, you can read in a new file to perform true crosscorrelation. The output is centered around the selected template.

(Q)RS detect permits inspection of the time-varying internal filter outputs in the QRS detection algorithm described in the book. This algorithm is designed for signals sampled at 200 sps. QRS detection operates on the entire file. When it encounters the end of the data file, it resets the threshold and internal data registers of the filters and starts over, so you may observe a "standing" wave if the data file is a short one.

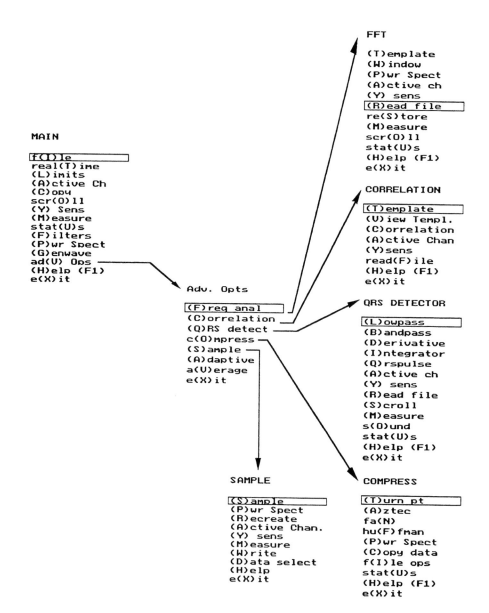

Figure D.5 Advanced options menus.

Function c(O)mpress provides the option to do data compression of a waveform using the turning point, AZTEC, Fan, and Huffman coding algorithms. The data reduction techniques compute approximations to the data in the upper channel. The algorithms operate only on the displayed data. The Turning Point algorithm reduces the number of data points by a factor of two, keeping critical points. The FAN and AZTEC algorithms require a threshold, which determines a trade-off between data reduction and distortion. The user is prompted to enter a value for the threshold, preferably a fraction of the data range (which is displayed). Huffman coding is a lossless algorithm that creates a lookup coding table based on frequency of occurrence of data values. A lookup table must be computed by (M)ake before the data can be compressed by (R)un. (R)un actually compresses then decompresses the data and displays the data reduction ratio. When making the table, first differencing can be used, which generally reduces the range of the data and improves data reduction. If the data range is too great when executing (M)ake, the range of the lookup table will be truncated to 8 bits, and values outside this range will be placed in the infrequent set and prefixed, so data reduction will be poor. The best thing to do in this case is to attenuate the data to a lower range (with a filter like atten40.fil) and then repeat the process.

(S)ample facilitates study of the sampling process by providing the ability to sample waveforms at different rates and reconstruct the waveforms using three different techniques. This module uses one of three waveforms which it generates internally, so you cannot use this module to subsample existing data. The data is generated at 5,000 sps, and you may choose a sampling rate from 1 to 2,500 sps at which to subsample. When you take the power spectrum of data that has been sampled (but not reconstructed), the program creates a temporary buffer filled with 512 points of the original waveform sampled at the specified rate. In other words, even though the displayed data is shown with the intersample spaces, the power spectrum is not computed based on the displayed data (and the 5,000 sps rate) but rather with the data that would have been created by sampling the analog waveform at the specified sample rate. The reason for this is to illustrate a frequency domain representation based on the specified sampling rate, and not the more complicated power spectrum that would result from computing the FFT of the actual displayed data.

Function (A)daptive demonstrates the basic principles of adaptive filtering, and a(V)erage illustrates the technique of time epoch signal averaging.

Figure D.6 shows the GENWAVE function that provides a waveform generator. Signals with controlled levels of ran(D)om and 60-H(Z) noise can be synthesized for testing filter designs. In addition to (S)ine, (T)riangle, and s(Q)uare waves, ECGs and other repetitive template-based waveforms can be generated with the t(E)mplate command. Figure D.7 shows two signals synthesized using this function. Figure D.8 shows the nine different templates that are provided for synthesizing normal and abnormal ECG signals. For more details about the GENWAVE function, see Appendix E.

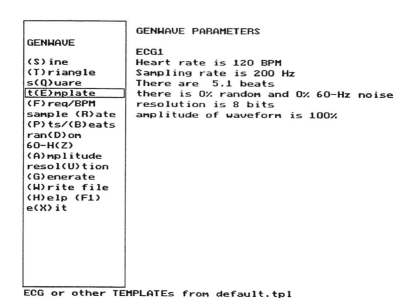

```
GENWAVE                      GENWAVE PARAMETERS

                             ECG1
(S)ine                       Heart rate is 120 BPM
(T)riangle                   Sampling rate is 200 Hz
s(Q)uare                     There are  5.1 beats
t(E)mplate                   there is 0% random and 0% 60-Hz noise
(F)req/BPM                   resolution is 8 bits
sample (R)ate                amplitude of waveform is 100%
(P)ts/(B)eats
ran(D)om
60-H(Z)
(A)mplitude
resol(U)tion
(G)enerate
(W)rite file
(H)elp (F1)
e(X)it
```

ECG or other TEMPLATEs from default.tpl

Figure D.6 Waveform generation.

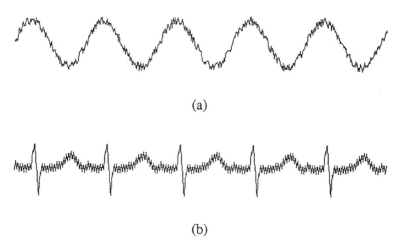

(a)

(b)

Figure D.7 Waveforms generated using (G)enwave. (a) Two-Hz sine wave with 20% random noise. (b) ECG based on ECG WAVE 1 with 5% 60-Hz noise.

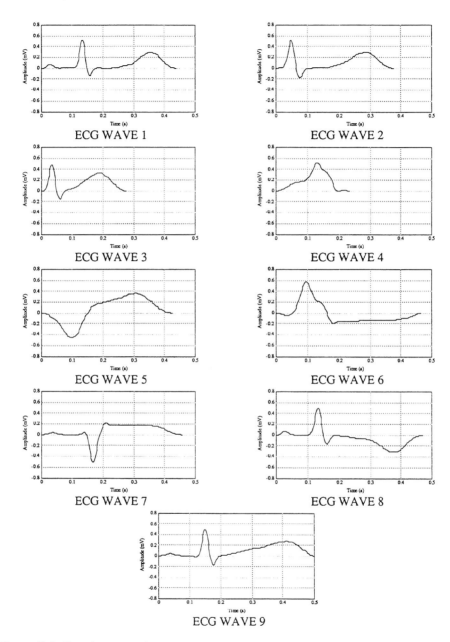

Figure D.8 Templates used by the GENWAVE function to generate ECG waveforms. Template ECG WAVE 1 is normal, the rest are abnormal.

Appendix E

Signal Generation

Thomas Y. Yen

In the process of designing digital filters and signal detection algorithms, it is important to have techniques for testing their performance. The testing process often involves applying signals to the input of the filter or algorithm and observing the resulting signal at the output. For biomedical applications, the use of physiological signals like the ECG and EEG are often preferable to basic signals like sine, triangle, and square waves. Physiological signals are not usually readily available making testing of the filters and algorithms difficult. It is, however, possible to use a computer and software to generate realistic physiological signals. This appendix describes how standard signals and physiological signals are generated by the GENWAVE function in UW DigiScope.

E.1 SIGNAL GENERATION METHODS

Two basic techniques for synthesizing signals are the model equation method and the waveform template method. In the model equation method, a mathematical model of the signal is used to calculate a sequence of values that make up the signal. This method of signal generation is very flexible, allowing signals of any frequency and amplitude to be produced. However, accurately modeling a particular signal such as the ECG can be difficult. If the signal is very complex, the model may require a large set of equations and hence a large computation time. We use the equation method in UW DigiScope for generating sine, square, and triangle waves.

On the other hand, the use of waveform templates is a very simple, flexible, and fast method of generating complex, repetitive signals. A waveform template is easily obtained by sampling one cycle of a signal with an analog-to-digital converter and saving the sampled data points as a template. This method allows any type of repetitive signal, regardless of complexity, to be reproduced. We used this template approach for ECG signal simulation in UW DigiScope because of its straightforward implementation and because of the ability to synthesize new signals by simply adding templates.

E.2 SIGNAL GENERATOR PROGRAM (GENWAVE)

GENWAVE is the program module embedded within UW DigiScope that is called with the (G)enwave command. It allows the creation of various types of signal waveforms. The user can specify the number of cycles, sampling rate, waveform repetition frequency, noise level, amplitude, and bit resolution. The different types of signals available are determined by the waveform templates available. New waveform templates can either be added to the default template file or saved in separate template files. The following sections describe how to create template files and how to use the GENWAVE program.

E.2.1 Creating template files

Two programs provided for the creation of templates are ASC2TPL.EXE and TPL2ASC.EXE. ASC2TPL.EXE converts an ASCII file of the proper format into a binary template file. TPL2ASC.EXE converts a binary template file, such as the default template file DEFAULT.TPL into an ASCII file so that it can be edited with an editor or word processing program. The default template file that comes with DigiScope includes nine ECG templates. New templates can be added to this file by first converting the file into an ASCII file, then adding new template data to the end of the file, and finally converting the ASCII file back to the binary default template format. In the conversion to binary, the ASC2TPL.EXE program automatically scales the waveforms to a 12-bit range to obtain maximal resolution of the waveform. Figure E.1 shows the structure of the ASCII template file format.

```
Waveform Name               (80 characters max.)
Waveform ID number          (1-12 are reserved; max. number 50)
Number of Datapoints        (32767 max.)
Sample Rate of Signal       (in Hz.-32767 max.)
Data(1)
Data(2)                     (the first and the last datapoints
 .                           must be the same baseline values)
 .
 .
Data(Number of Datapoints)
(Repeat above for each waveform template)
```

Figure E.1 ASCII template file format.

To convert the default template file to an ASCII template file, enter the following syntax:

`TPL2ASC DEFAULT.TPL <ASCII File Name>`

where `ASCII File Name` is the name by which the default template data is stored. The file name `DEFAULT.TPL` is the binary template file from which the waveform templates are read. An example of the format of this command is

`TPL2ASC DEFAULT.TPL ASCTEMP.ASC`

This command converts the binary template file `DEFAULT.TPL` into an ASCII template file called `ASCTEMP.ASC`. Both file names must be specified, and a different binary template file name other than `DEFAULT.TPL` may be used.

To convert an ASCII template file to the binary default template file, first rename the current default template file, then type the following at the command prompt:

`ASC2TPL <ASCII File Name> DEFAULT.TPL`

where `ASCII File Name` is the file that is to be converted to the default template file. The new `DEFAULT.TPL` file must then be placed in the `DIGSCOPE\STDLIB` directory in order for DigiScope to use it. An example of the format of this command is

`ASC2TPL ASCTEMP.ASC DEFAULT.TPL`

converts the ASCII template file into the template file called `DEFAULT.TPL`. Both file names must be specified, and a different binary template file name other than `DEFAULT.TPL` may be used.

A template file can have a maximum of 50 waveform templates. Each waveform must have a waveform ID. The ID numbers in a given file should start with one and increase by one for each template in the file. The waveform name can be up to 80 characters in length. To ensure amplitude-matching of the start and the end of a template, the first and last waveform values must be the same. Figure E.2 shows an example of an ASCII template file.

E.2.2 Using the GENWAVE function

When the signal generator `GENWAVE` creates a waveform, several parameters are attached. If the signal is saved as a file, these parameters are placed in the file header (defaults are in brackets).

1. Output filename (8 characters with 3 letter extension) [`WAVEFORM.OUT`]
2. Input filename (8 characters with 3 letter extension) [`DEFAULT.TPL`]
3. Output sampling rate in sps (10,000 max.) [500]
4. Output waveform frequency in cycles per minute (10,000 max.) [60]

5. Number of data points (5,000 max.) [512]
6. Noise: % of full scale (0 to ±100) [0]
   (a) Random noise
   (b) 60 Hz noise
7. Waveform ID (1 to 50) [1]
8. Full-scale resolution in bits (3–12 max.) [8]

GENWAVE is always able to produce sine, triangle, and square wave signals because they are built into the program even if the DEFAULT.TPL file is not found. Since GENWAVE always reads this file if it is present, should you wish to create a new default file, save the old DEFAULT.TPL file by another name and rename your new file DEFAULT.TPL.

```
Normal ECG              (Waveform Name)
4                       (Waveform ID)
157                     (Number of Datapoints)
360                     (Samplerate)
0                       (Data1)
0                       (Data2)
0                       (Data3)
624                     (Data4)
1560                    (Data5)
.                       .
.                       .
.                       .
920                     (Data155)
25                      (Data156)
0                       (Data157 Last)
PVC                     (Waveform Name)
5                       (Waveform ID)
83                      (Number of Datapoints)
360                     (Samplerate)
0                       (Data1)
0                       (Data2)
43                      (Data3)
324                     (Data4)
260                     (Data5)
.                       .
.                       .
.                       .
320                     (Data1)
35                      (Data82)
0                       (Data83 Last)
```

Figure E.2 Example of an ASCII template file.

Appendix F

Finite-Length Register Effects

Steven Tang

Digital filter designs are either implemented using specific-purpose hardware or general-purpose computers such as a PC. Both approaches involve the use of finite-length data registers (FLR) which can represent only a limited number of values. In Chapter 3, we discussed the process of analog-to-digital conversion and some of the quantization effects due to the finite number of representable magnitudes. This is one of several problems dealing with FLR effects that we must consider when designing our own digital filter on a signal processor or microprocessor. This appendix discusses overflow characteristics, roundoff noise, limit cycles, scaling and I/O variations due to fixed-point registers. We also compare the functional advantages of floating-point and fixed-point registers.

F.1 QUANTIZATION NOISE

Fixed-point registers, when used in digital filters, store a finite number of representable integer numbers. There are two consequences of this type of representation: (1) the state variables that make up the filter can only represent an integral multiple of the smallest quantum, and (2) there is a maximal value that the register can represent in a one-to-one correspondence. The first effect is known as quantization, the second, as overflow (Oppenheim and Schafer, 1975; Roberts and Mullis, 1987).

Quantization errors can occur in a fixed-point register whenever there is a multiply and accumulate function. For example, if we are trying to implement an FIR filter, the output would be the sum of weighted tapped inputs. The final summation is representable only to the precision of the smallest value. The smallest quantum that we can represent with a register of bit-length $B+1$ is

$$q = \Delta 2^{-B} \qquad (F.1)$$

This value q is frequently called the quantization step size. We can arbitrarily define q by choosing the value of Δ. If we are trying to design an integer filter, the value of Δ is 2^B, and $q = 1$. The advantage of choosing a large Δ is that we can expand the domain of representable values. However, we consequently lose precision since we also increase the quantization step size. Figure F.1 shows the quantizer characteristics for the rounding of a three-bit register.

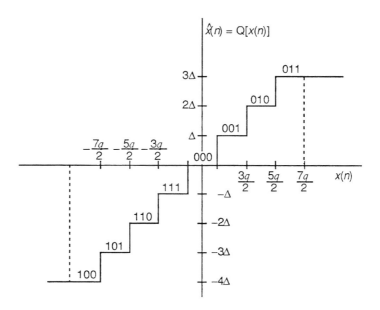

Figure F.1 2's complement rounding quantization effect for 3-bit register.

The error between the real number $x(n)$ and its finite binary representation is

$$e = x(n) - Q[x(n)] \tag{F.2}$$

This is known as the quantization error due to FLR. This presumably also occurs during data acquisition with any A/D converter. However, A/D quantization is a hardware problem; FLR quantization is due to software restrictions.

F.1.1 Rounding

One method of implementing a quantizer is to round off the true value to the nearest representable quantum level. The quantization error is then bounded by $q/2$ and $-q/2$. Thus, quantization is functionally equivalent to adding some random noise in the range $-q/2 < \alpha < q/2$ to the original real number. We can think of this noise as being randomly generated and having the probably density function (PDF) shown in Figure F.2.

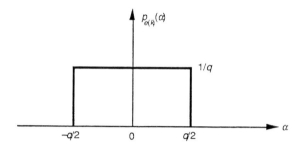

Figure F.2 The probability density function of quantization error for rounding.

Theoretical studies and numerical simulations have shown that this model is accurate for inputs with a relatively broad frequency spectrum and wide PDF, so that the input does not remain between the same quantization levels for a long duration. We can thus calculate a theoretical value for noise power associated with each quantized multiply and accumulate function.

$$\sigma_e{}^2 = E[e(k)^2] = \int_{-q/2}^{q/2} \frac{\alpha^2}{q} d\alpha = \frac{q^2}{12} \tag{F.3}$$

F.1.2 Truncation

The other method of quantizing is to simply truncate the value, or reduce it until we find a representable level. This has a similar effect to rounding except that the probability density function of such quantization error has a shifted mean. The noise range has limits of $-q$ and 0. While there is no inherent advantage to using either rounding or truncation, rounding is preferred because of its zero mean. For theoretical calculations, this is a much easier method to use since the noise power is simply the variance of the noise.

F.2 LIMIT CYCLES

If a digital filter amplifies the input, there will be internal gain among the state variables. If the accumulated value at a register is beyond the highest binary representation available, an overflow characteristic must be implemented. Figure F.3 shows several different ways of describing a function for the values outside the representable range. Saturation overflow acts as a strict output limiter. Two's complement overflow has the characteristic of a wraparound effect. Zero after overflow simply suppresses the output to zero.

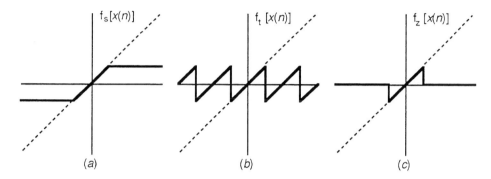

Figure F.3 Overflow characteristics. (a) Saturation, (b) Two's complement, (c) Zero after overflow.

In each case, there are competing advantages that offer a compromise between total roundoff error and the likelihood of continual overflow. Saturation overflow provides the closest output to the actual value (shown by the dotted line). However, by maintaining the output at the largest quantized level, there is greater potential for another overflow after the next filter iteration. The two's-complement overflow scheme is a natural characteristic of two's complement representation; the largest positive representable value is one bit smaller than the most negative value available. While two's-complement creates greater roundoff error, there is equal probability that the output will fall anywhere within the representable quantization range. The zero after overflow is a compromise between the saturation and two's-complement characteristics.

F.2.1 Overflow oscillations

From the different types of overflow characteristics, we see that the possibility of repeated overflows can happen in all three cases. When overflows occur continually, and the output does not converge to zero after an initial nonzero input, there exists an overflow oscillation. The implication of an overflow oscillation is that the output is no longer dependent on the input.

Figure F.4 shows a second-order filter example of jumping from an allowable state space to an overflowed state, applying two's-complement overflow, and finally using roundoff quantization. For this example, we simplify the typical state space equations by setting the input to be zero. We can represent any filter by the equations

$$x(k + 1) = Ax(k) + Bu(k) \tag{F.4}$$

$$y(k) = Cx(k) + Du(k) \tag{F.5}$$

Thus, the A is the matrix that maps the current state variable vector, $x(k)$, to the next, $x(k + 1)$. If the eigenvalues of A [i.e., the poles of $H(z)$] are greater than unity, the filter is unstable and overflow will occur. There are particular regions in the state space for which $x(k + 1)$ will continually be mapped out of range for zero input. These are the areas for which overflow oscillations occur. Having a nonzero input can sometimes change the characteristics of the mapping function so that it jumps out of this cycle. However, the input is usually not very large in comparison to the total range of the state space.

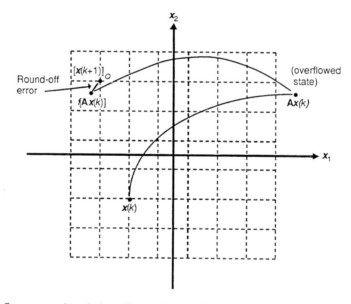

Figure F.4 State space description of second-order digital filter showing overflow and roundoff.

F.2.2 Deadband effect

Overflow is not the only type of limit cycle that can occur from FLR effects. Recursive IIR filters with constant input can produce steady-state output after quantization when they should actually continually decay. This problem is a result of roundoff errors that cause the state variable vector to get stuck in the same state space. Thus, there is essentially a *deadband* where constant input prevents a linear I/O mapping of the filter characteristics. This is not as hazardous as overflow oscillation in the amount of roundoff error that occurs, but it still is problematic in giving an erroneous output response.

Here is a first order example of such a filter. Let

$$y(nT) = -0.96y(nT - T) + x(nT) \qquad \text{for } x(0) = 13 \text{ and } x(nT) = 0 \text{ for } n > 0$$

Clearly, without quantization, the output should eventually decay to zero. However, if we assume rounding to the nearest integer at the output, we will have the following I/O characteristics.

n	$y(nT) = -0.96[y(nT - T)]_Q$	$[y(nT)]_Q$
0	13	13
1	−12.48	−12
2	11.52	12
3	−11.52	−12
4	11.52	12

Such a filter is said to have a limit cycle period of two. A filter with limit cycle of periodicity one would have a state vector that remains inside the same grid location continuously.

One solution to deadband effects is to add small amounts of white noise to the state vector $x(nT)$. However, this also means that the true steady state of a filter response will never be achieved. Another method is to use magnitude truncation instead of rounding. The problem here, as well, is that truncation may introduce new deadbands while eliminating the old ones.

F.3 SCALING

There are several ways to avoid the disastrous effect of limit cycles. One is to increase Δ so that the state space grid covers a greater domain. However, this also increases the quantization noise power. The most usual cause of limit cycles are filters that have too much gain. Scaling is a method by which we can reduce the chances of overflow, still maintain the same filter transfer function, and not compromise in limiting quantization noise.

Figure F.5 shows how to scale a node v' so that it does not overflow. By dividing the transfer function $F(z)$ from the input u to the node by some constant β, we scale the node and reduce the probability of overflow. To maintain the I/O characteristics of the overall filter, we must then add a gain of β to the subsequent transfer function $G(z)$.

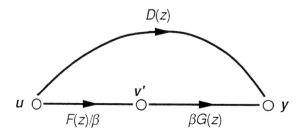

Figure F.5 State variable description of scale node variable v'.

There are two different scaling rules called l_1 and l_2 scaling. The first rule corresponds to bounding the absolute value of the input. The second rule maintains a bounded energy input. The two rules are

$$\beta = \|f\|_1 = \sum_{l=0}^{\infty} |f(l)| \tag{F.6}$$

$$\beta = \delta \|f\|_2 = \delta \left[\sum_{l=0}^{\infty} f^2(l) \right]^{1/2} \tag{F.7}$$

The parameter δ can be chosen arbitrarily to meet the desired requirements of the filter. It can be regarded as the number of standard deviations representable in the node v' if the input is zero mean and unit variance. A δ of five would be considered very conservative.

F.4 ROUNDOFF NOISE IN IIR FILTERS

Both roundoff errors and quantization errors get carried along in the state variables of IIR filters. The accumulated effect at the output is called the roundoff noise. We can theoretically estimate this total effect by modeling each roundoff error as an additive white noise source of variance $q^2/12$.

If the unit-pulse response sequence from node i to the output is g_i, and if quantization is performed after the accumulation of products in a double length accumulator, the total output roundoff noise is estimated as

$$\sigma^2_{\text{tot}} = \frac{q^2}{12} \left\{ \sum_{i=1}^{n} \|g_i\|^2 + 1 \right\} \tag{F.8}$$

The summation in this equation is sometimes called the noise gain of the filter. Choosing different forms of filter construction can improve noise gain by as much as two orders of magnitude. Direct form filters tend to give higher noise gains than minimum noise filters that use appropriate scaling and changes to g_i to reduce the amount of roundoff noise.

F.5 FLOATING-POINT REGISTER FILTERS

Floating-point registers are limited in their number of representable states, although they offer a wider domain because of the exponential capabilities. The state space grid no longer looks uniform but has a dependency on the distance between the state vector and the origin. Figure F.6 demonstrates the wider margins between allowable states for numbers utilizing the exponent range.

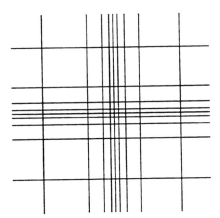

Figure F.6 State space grid for floating-point register.

Floating point differs from fixed point in two ways: (1) there is quantization error present during addition, and (2) the output roundoff noise variance is proportional to the size of the number in the state variable. Although floating point greatly expands the domain of a filter input, the accumulated roundoff errors due to quantization are considerably greater than for fixed-point registers.

F.6 SUMMARY

In many real-time applications, digital signal processing requires the use of FLRs. We have summarized the types of effects and errors that arise as a result of using FLRs. These effects depend on the type of rounding and overflow characteristics of a register, whether or not it is fixed or floating point, and if there is scaling of the internal nodes.

Figure F.7 compares the total error for a filter with variable scaling levels. For no scaling, we expect to have greater probability of overflow, unless the input is well bounded. As we increase the scaling factor δ, overflow is less prevalent, but the roundoff error from quantization begins to increase because the dynamic range of the node register is decreased. To minimize total error output, we must find a compromise that decreases both errors.

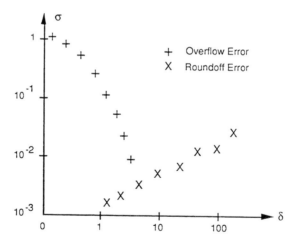

Figure F.7 Comparison of overflow and roundoff error in total error output.

F.7 LAB: FINITE-LENGTH REGISTER EFFECTS IN DIGITAL FILTERS

Write a subroutine that quantizes using the rounding feature and a subroutine that simulates 2's-complement overflow characteristics for an 8-bit integer register. Try implementing a high-pass IIR filter with the transfer function

$$H(z) = \frac{1}{1 + 0.5z^{-1} + z^{-2}}$$

Using a sinusoidal input, find the amplitude at which the filter begins to overflow. Examine the output of the filter for such an input. Does the overflow characterize a 2's-complement response?

F.8 REFERENCES

Oppenheim, A. V. and Schafer, R. W. 1975. *Digital Signal Processing*. Englewood Cliffs, NJ: Prentice Hall.

Roberts, R. A. and Mullis, C. T. 1987. *Digital Signal Processing*, Reading, MA: Addison-Wesley.

Appendix G

Commercial DSP Systems

Annie Foong

A wide variety of commercial data acquisition hardware and software is currently available in the market. Most comes in the form of full-fledged data acquisition systems that support various hardware cards in addition to data analysis and display capabilities. Basically a complete data acquisition system consists of three modules: acquisition, analysis, and presentation.

G.1 DATA ACQUISITION SYSTEMS

G.1.1 Acquisition

Four common ways of acquiring data use (1) an RS-232 serial interface, (2) the IEEE 488 (GPIB) parallel instrumentation interface, (3) the VXI bus, or (4) a PC-bus plug-in data acquisition card.

RS-232 interface

This approach consists of a serial communication protocol for simple instruments such as digital thermometers, panel meters, and data loggers. They are useful for controlling remote data acquisition systems from long distances at data rates lower than 1 kbyte/s. Since the RS-232 interface comes standard on most computers, no extra hardware is necessary.

IEEE 488 (GPIB) interface

Many sophisticated laboratory and industrial instruments, such as data loggers and digital oscilloscopes, are equipped with GPIB interfaces. Devices communicate through cables up to a maximum length of 20 meters using an 8-bit parallel protocol with a maximum data transfer rate of two Mbyte/s. This interface supports both control and data acquisition. IEEE 488 uses an ASCII command set (Baran, 1991).

VXI bus

This bus is a high-performance *instrument-on-a-card* architecture for sophisticated instruments. Introduced in 1987, this architecture has been driven by the need for physical size reduction of rack-and-stack instrumentation systems, tighter timing and synchronization between multiple instruments, and faster transfer rates. This standard is capable of high transfer speeds exceeding 10 Mbyte/s.

Plug-in data acquisition boards

Data acquisition boards plug directly into a specific computer type, such as the PC or the Macintosh. This method combines low cost with moderate performance. These boards usually support a wide variety of functions including A/D conversion, D/A conversion, digital I/O, and timer operations. They come in 8–16 bit resolution with sampling rates of up to about 1 MHz. They offer flexibility and are ideal for general-purpose data acquisition.

G.1.2 Analysis and presentation

Data analysis transforms raw data into useful information. This book is principally about data analysis. Most software packages provide such routines as digital signal processing, statistical analysis, and curve fitting operations.

Data presentation provides data to the user in an intuitive and meaningful format. In addition to presenting data using graphics, presentation also includes recording data on strip charts and generation of meaningful reports on a wide range of printers and plotters.

G.2 DSP SOFTWARE

The trend is toward using commercial DSP software that provides the entire process of data acquisition, analysis, and presentation. Here we discuss commercially available software for general plug-in PC data acquisition boards. Because of the flexibility of such a scheme of data acquisition, there is a huge market and many suppliers for such software. In addition, many vendors offer complete training programs for their software.

Software capabilities vary with vendors' emphasis and pricing. Some companies, for example, sell their software in modules, and the user can opt to buy whatever is needed.

Some common capabilities of commercial DSP software include the following:

1. Support of a wide variety of signal conversion boards.
2. Comprehensive library of DSP algorithms including FFT, convolution, low-pass, high-pass, and bandpass filters.
3. Data archiving abilities. The more sophisticated software allows exporting data to Lotus 123, dBase, and other common analysis programs.
4. Wide range of sampling rates.
5. Impressive graphics displays and menu and/or icon driven user interface.
6. User-programmable routines.
7. Support of high-level programming in C, BASIC, or ASCII commands.
8. Customizable report generation and graphing (e.g., color control, automatic or manual scaling).

A few interesting software packages are highlighted here to give the reader a flavor of what commercial DSP software offers.

SPD by Tektronix is a software package designed for Tektronix digitizers and digital oscilloscopes and the PEP series of system controllers or PC controllers. It offers in its toolset over 200 functions including integration and differentiation, pulse measurements, statistics, windowing, convolution and correlation, forward and inverse FFTs for arbitrary length arrays, sine wave components of an arbitrary waveform, interpolation and decimation, standard waveform generation (sine, square, sinc, random), and FIR filter generation.

DADiSP by DSP Development Corporation offers a version that operates in the protected mode of Intel 80286 or 80386 microprocessors, giving access to a full 16 Mbytes of addressability. Of interest is the metaphor that DADiSP uses. It is viewed as an interactive graphics spreadsheet. The spreadsheet is for waveforms, signals, or graphs instead of single numbers. Each cell is represented by a window containing entire waveforms. For example, if window 1 (W1) contains a signal, and W2 contains the formula DIFF(W1) (differentiate with respect to time), the differentiated signal will then be displayed in W2. If the signal in W1 changes, DADiSP automatically recalculates the derivative and displays it in W2. It also takes care of assigning and managing units of measurement. In the given example, if W1 is a voltage measurement, W1 will be rendered in volts, and W2 in volts per second. As many as 100 windows are allowed with zoom, scroll, and cursoring abilities. The number of data points in any series is limited only by disk space, as DADiSP automatically pages data between disk and memory.

DspHq by Bitware Research Systems is a simple, down-to-earth package that includes interfaces to popular libraries such as MathPak87 and Numerical Recipes.

MathCAD by MATHSoft, Inc. is a general software tool for numerical analysis. Although not exactly a DSP package, its application packs in electrical engineering and advanced math offer the ability to design IIR filters, perform convolution and correlation of sequences, the DFT in two dimensions, and other digital filtering.

A more powerful software package, MatLAB by Math Works, Inc., is also a numerical package, with an add-on Signal Processing Toolbox package having a rich collection of functions immediately useful for signal processing. The Toolbox's features include the ability to analyze and implement filters using both direct and FFT-based frequency domain techniques. Its IIR filter design module allows the user to convert classic analog Butterworth, Chebyshev, and elliptic filters to their digital equivalents. It also gives the ability to design directly in the digital domain. In particular, a function called *yulewalk*() allows a filter to be designed to match any arbitrarily shaped, multiband, frequency response. Other Toolbox functions include FIR filter design, FFT processing, power spectrum analysis, correlation function estimates and 2D convolution, FFT, and crosscorrelation. A version of this product limited to 32×32 matrix sizes can be obtained inexpensively for either the PC or Macintosh as part of a book-disk package (*Student Edition of MatLAB*, Prentice Hall, 1992, about $50.00).

ASYST by Asyst Software Technologies supports A/D and D/A conversion, digital I/O, and RS-232 and GPIB instrument interfacing with a single package. Commands are hardware independent. It is multitasking and allows real-time storage to disk, making it useful for acquiring large amounts of data at high speeds.

The OMEGA SWD-RTM is a real-time multitasking system that allows up to 16 independent timers and disk files. This is probably more useful in a control environment that requires stringent timing and real-time capabilities than for DSP applications.

LabWindows and LabVIEW are offered by National Instruments for the PC and Macintosh, respectively. LabWindows provides many features similar to those mentioned earlier. However, of particular interest is LabVIEW, a visual programming language, which uses the concept of a virtual instrument. A virtual instrument is a software function packaged graphically to have the look and feel of a physical instrument. The screen looks like the front panel of an instrument with knobs, slides, and switches. LabVIEW provides a library of controls and indicators for users to create and customize the look of the front panel. LabVIEW programs are composed of sets of graphical functional blocks with interconnecting wiring. Both the virtual instrument interface and block diagram programming attempt to shield engineers and scientists from the syntactical details of conventional computer software.

G.3 VENDORS

Real Time Devices, Inc.
   State College, PA
   (814) 234-8087

BitWare Research Systems
   Inner Harbor Center, 8th Floor, 400 East Pratt Street,
   Baltimore, MD 21202-3128
   (800) 848-0436

Asyst Software Technologies
   100 Corporate Woods, Rochester, NY 14623
   (800) 348-0033

National Instruments
   6504 Bridge Point Parkway, Austin, TX 78730-5039
   (800) IEEE-488

Omega Technologies
   One Omega Drive, Box 4047, Stamford, CT 06907
   (800) 826-6342

DSP Development Corporation
   One Kendall Square, Cambridge, MA 02139
   (617) 577-1133

Tektronix
   P.O. Box 500, Beaverton, OR 97077
   (800) 835-9433

MathSoft, Inc.
   201 Broadway Cambridge, MA 02139
   (800) MathCAD

The MathWorks, Inc.
   21 Eliot St, South Natick, MA 01760
   (508) 653-1415

G.4 REFERENCES

Baran, N. 1991. Data acquisition: PCs on the bench. *Byte*, 145–49, (May).
Coffee, P. C. 1990. Tools provide complex numerical analysis. *PC Week*, (Oct. 8).
National Instruments. 1991. IEEE-488 and VXI bus control. *Data Acquisition and Analysis*.
Omega Technologies 1991. *The Data Acquisition Systems Handbook*.
The MathWorks, Inc. 1988. *Signal Processing Toolbox User's Guide*.

Index